Reliability and Warranties:
Methods for Product Development and Quality Improvement

RELIABILITY AND WARRANTIES:
METHODS FOR PRODUCT DEVELOPMENT AND QUALITY IMPROVEMENT

MARLIN U. THOMAS
AIR FORCE INSTITUTE OF TECHNOLOGY
WRIGHT-PATTERSON AIR FORCE BASE, OHIO, USA

Taylor & Francis
Taylor & Francis Group
Boca Raton   London   New York

A CRC title, part of the Taylor & Francis imprint, a member of the Taylor & Francis Group, the academic division of T&F Informa plc.

Published in 2006 by
CRC Press
Taylor & Francis Group
6000 Broken Sound Parkway NW, Suite 300
Boca Raton, FL 33487-2742

© 2006 by Taylor & Francis Group, LLC
CRC Press is an imprint of Taylor & Francis Group

No claim to original U.S. Government works
Printed in the United States of America on acid-free paper
10 9 8 7 6 5 4 3 2 1

International Standard Book Number-10: 0-8493-7149-X (Hardcover)
International Standard Book Number-13: 978-0-8493-7149-3 (Hardcover)
Library of Congress Card Number 2005055983

This book contains information obtained from authentic and highly regarded sources. Reprinted material is quoted with permission, and sources are indicated. A wide variety of references are listed. Reasonable efforts have been made to publish reliable data and information, but the author and the publisher cannot assume responsibility for the validity of all materials or for the consequences of their use.

No part of this book may be reprinted, reproduced, transmitted, or utilized in any form by any electronic, mechanical, or other means, now known or hereafter invented, including photocopying, microfilming, and recording, or in any information storage or retrieval system, without written permission from the publishers.

For permission to photocopy or use material electronically from this work, please access www.copyright.com (http://www.copyright.com/) or contact the Copyright Clearance Center, Inc. (CCC) 222 Rosewood Drive, Danvers, MA 01923, 978-750-8400. CCC is a not-for-profit organization that provides licenses and registration for a variety of users. For organizations that have been granted a photocopy license by the CCC, a separate system of payment has been arranged.

Trademark Notice: Product or corporate names may be trademarks or registered trademarks, and are used only for identification and explanation without intent to infringe.

Library of Congress Cataloging-in-Publication Data

Thomas, Marlin U., 1942-
  Reliability and warranties : methods for product development and quality improvement / Marlin U. Thomas.
       p. cm.
  Includes bibliographical references and index.
  ISBN 0-8493-7149-X
  1. New products. 2. Production management. I. Title.

TS 170.T48 2006
658.5'75--dc22   2005055983

Taylor & Francis Group
is the Academic Division of Informa plc.

Visit the Taylor & Francis Web site at
http://www.taylorandfrancis.com

and the CRC Press Web site at
http://www.crcpress.com

Preface

Product quality is the collective influence of several elements that affect how well a product is made, how it performs, and how customers accept it. Reliability is perhaps the single most vital of these elements because it represents the failure-free service that a customer receives from the product. Reliability also has a major impact on the numbers and costs of warranty claims and related measures of customer satisfaction. Consequently, understanding the relationship between quality, reliability, and warranty is essential for ensuring that products are produced with high quality. This level of understanding is also essential in analyzing and evaluating decisions involving warranty policies and strategies for improving the quality and reliability of products.

Warranties have existed for generations, but it has only been during the past 30 to 35 years that a literature has emerged on quantitative methods for treating them. Traditional textbooks on reliability and reliability engineering do not cover warranties. Two exceptions are books by Elsayed (1996) that devoted a chapter to warranty models, and Blischke and Murthy (2000) that included a chapter on cost models for warranties and service contracts. Most practitioners engaged today in product design, development, and marketing received little or no exposure to warranties as an integral part of their education and training in quality and reliability. There is need for bridging this gap and this book attempts to address this need.

The motivation for developing a book on quality, reliability, and warranty came from suggestions by numerous reliability students enrolled in the Purdue University Continuing Engineering Education program as well as attendees of tutorial sessions on quality and warranties at the annual Reliability and Maintenance Symposium. The purpose of this book is to provide a modern concept of product quality that integrates reliability and warranties. The level of the material presented is aimed at undergraduate seniors and first-year graduate students in engineering and operations research. The book is intended for use in a continuing engineering education short course in reliability, and as a supplementary text for courses in reliability and quality management. It is also suitable as a reference for practicing engineers and quality managers engaged in the design, development, and marketing of consumer products.

The topics covered in this book serve to expand the concept of quality as an overall product value to customers. Methods and guidelines are presented that address the following issues relative to a manufacturer or producer's point of view:

1. Analyses and characteristics of product failures
2. Assessing product reliability

3. Estimating and predicting warranty costs
4. Developing and evaluating quality improvement strategies

The author is grateful to the many students and colleagues who have contributed to the development of this book and especially to former associates at Chrysler (now DaimlerChrysler) Corporation where he was first introduced to warranty issues in vehicle reliability; General Motors and its support of the Purdue University Continuing Engineering Education program; and industrial engineering graduate student Yue Pan, who assisted in developing materials for numerous classes and presentations. Final acknowledgment goes to my good friend and colleague, the late Dr. William A. Golomski, whose encouragement led to this book.

Marlin U. Thomas
Air Force Institute of Technology

References

Blischke, W.R. and D.N.P. Murthy, 2000, *Reliablity—Modeling, Prediction, and Optimization*, Wiley and Sons, New York.

Elsayed, E.A., 1996, *Reliability Engineering*, Addison-Wesley, Boston.

The Author

Marlin U. Thomas, Ph.D., is dean, Graduate School of Engineering and Management at the Air Force Institute of Technology. Prior to joining AFIT he was professor and head of the School of Industrial Engineering at Purdue University where he had also served as director of the Institute for Interdisciplinary Engineering Studies. He received his B.S.E. (1967) from the University of Michigan Dearborn, and M.S.E. (1968) and Ph.D. (1971) in industrial engineering from the University of Michigan. He has been a registered professional engineer since 1970, with engineering R&D and product development experience with Owens-Illinois, Inc., and as a manager of vehicle reliability planning and analysis with Chrysler Corporation, where he first became interested in developing quantitative methods for analyzing warranties. He also has served as a program director for the National Science Foundation working in operations research and production systems.

Dr. Thomas' teaching and research areas include stochastic modeling, reliability, and logistics systems, with emphasis on contingency operations. He has received several awards for his professional accomplishments, including the Frank Groseclose Medallion Award (for his notable impact on the industrial engineering profession) and the Albert G. Holzman Distinguished Educator Award from the Institute of Industrial Engineers. He has authored or co-authored over 60 articles and technical papers that have appeared in publications such as *IIE Transactions, Operations Research, Management Science, IEEE Transactions on Reliability*, and he has delivered over 100 presentations and lectures at major conferences and universities.

Dr. Thomas has held the position of president of the Institute of Industrial Engineers; department editor, *IIE Transactions*; area editor, *Operations Research*; and was a member of the Army Science Board. He is a member of the American Society for Engineering Education and American Indian Science and Engineering Society; as well as a Fellow of the American Society for Quality, the Institute of Industrial Engineers, and the Institute for Operations Research and Management Sciences. He is also a Captain, Civil Engineer Corps, U.S. Navy Reserve (retired).

About This Book

Product quality is the collective influence of several elements that affect how well a product is made, how it performs, and how customers accept it. Reliability is perhaps the single most important element because it represents the amount of failure-free service that a customer receives from the product. Reliability also has a major impact on the number and costs of warranty claims and related measures of customer satisfaction. Consequently, in this book the author integrates an over-arching philosophy for product quality, reliability, and warranty based on the following principles:

1. Quality is a vector of attributes that relate to the way a product is designed, developed, produced, and accepted by customers.
2. Reliability is a vital element of quality that impacts all other elements.
3. Warranty feedback provides a relative measure of product performance.

The book begins with preliminary results for analyzing failures, which includes the common distributions, application of probability transforms, and methods for fitting failure time distributions. The topics covered serve to expand the concept of quality as an overall product value to customers. Methods and guidelines are presented relative to a manufacturer or producer's point of view that address the following:

1. Analysis and characteristics of product failures
2. Assessing product reliability
3. Estimating and predicting warranty costs
4. Developing and evaluating quality improvement strategies

A unique feature of this book is a framework presented in Chapter 5 for measuring and tracking overall quality performance. To maintain high quality it is important to understand the causes of quality problems, to be able to track performance over time, and to have some capability of providing responsive actions for improvements. Within the book, relative quality indicators are defined as well as methods for tracking them by the month of production and month in service, using MOP/MIS charts along with examples. The book also includes an approach for establishing product quality improvement targets that includes a decision analysis framework for structuring improvement options, identifying failure modes and criticality analysis, and methods for allocating lower subsystem or element improvement goals. Examples are used throughout the book to illustrate results, and exercises are provided for reinforcing key topics.

Marlin U. Thomas

Contents

Chapter 1 Product Quality and Warranty..1
1.1 Introduction..1
1.2 Background and Scope...2
References...5

**Chapter 2 Preliminary Results for Analyzing
Product Failures**...7
2.1 Failure Time Distributions...7
    2.1.1 Some Common Continuous Distributions..................8
    2.1.2 Laplace Transforms..11
    2.1.3 Some Common Discrete Distributions.....................15
    2.1.4 Probability Generating Functions.............................16
2.2 Fitting Distributions...21
    2.2.1 Goodness of Fit Testing...22
    2.2.2 Probability Plotting..24
2.3 Counting the Number of Failures....................................26
    2.3.1 Binomial Failure Process...27
    2.3.2 Poisson Failure Process...29
    2.3.3 Approximations to M(t)..30
2.4 Exercises..32
References..34

Chapter 3 Quality and Reliability..35
3.1 Quality Concepts..35
    3.1.1 Product Life and Quality...37
    3.1.2 The Product System...37
3.2 Product Reliability..38
    3.2.1 Reliability Measures..38
        3.2.1.1 Exponential Failure Times..........................41
        3.2.1.2 Weibull Failure Times.................................42
    3.2.2 Failures and Lifetime Characteristics......................44
        3.2.2.1 Infant Mortality Period...............................45
        3.2.2.2 Useful Life Period.......................................45
        3.2.2.3 Wear-out Period..45
    3.2.3 System Configurations..46
        3.2.3.1 Series System...46
        3.2.3.2 Parallel System..46
        3.2.3.3 Combined Series/Parallel System............47

        3.2.4   Reliability Improvement and Redundancy 50
3.3   Load and Capacity Models 51
        3.3.1   Periodic Loadings 54
        3.3.2   Random Loadings 55
3.4   Exercises 56
References 58

Chapter 4   Economic Models for Product Warranties **59**
4.1   Definitions and Types of Warranties 59
        4.1.1   Free-Replacement Warranty (FRW) 60
        4.1.2   Pro Rata Replacement Warranty (PRW) 61
        4.1.3   Warranty Costs and Discounting 62
        4.1.4   Combined FRW/PRW 64
4.2   Warranty Cost Models 66
        4.2.1   Nonrenewing Warranty 67
                4.2.1.1   FRW Policy 68
                4.2.1.2   PRW and Combined Policies 72
        4.2.2   Renewing Warranty 72
4.3   Determining Optimum Warranty Periods 74
        4.3.1   Exponential Failure Times 76
        4.3.2   Weibull Failure Times 76
4.4   Estimating Warranty Reserve 77
        4.4.1   Exponential Failure Times 78
        4.4.2   Weibull Failure Times 79
4.5   Exercises 80
References 82

Chapter 5   Product Quality Monitoring and Feedback **85**
5.1   Quality System Monitoring and Control 85
        5.1.1   Features for an Effective TQC System 88
5.2   Multiattribute Quality Assessment 90
        5.2.1   An Approach for Assessing Overall Product Quality 91
        5.2.2   Relative Quality Indicators 94
5.3   Warranty Information Feedback Models 95
        5.3.1   The Claims Process 95
        5.3.2   Poisson Warranty Claims 96
        5.3.3   Poisson Warranty Claims with Lags 101
5.4   MOP/MIS Charts 104
        5.4.1   Average Defects per Unit Sold (DPUS) 107
        5.4.2   Average Warranty Cost per Unit Sold (CPUS) 108
        5.4.3   MOP/MIS Charts with Lag Factors 108
                5.4.3.1   Lag Factor Model 109
5.5   Exercises 111
References 114

Chapter 6 The Quality Improvement Process 115
6.1 Introduction 115
6.2 Identifying Quality Problems 116
   6.2.1 FMECA Process 116
   6.2.2 Fault Tree Analysis 119
      6.2.2.1 Procedure for Generating Cut Sets 121
6.3 Developing Quality Improvement Goals 125
   6.3.1 Thomas–Richard Method 127
      6.3.1.1 Constant Failure Rate (CFR) Allocation 128
      6.3.1.2 Increasing Failure Rate (IFR) Allocation 131
6.4 Decision Analysis Framework 133
   6.4.1 Decisions under Risk Conditions 134
   6.4.2 A Generalized Maximum Entropy Principle (GMEP) 138
   6.4.3 Decisions under Complete Uncertainty 140
6.5 Exercises 141
References 144

Chapter 7 Toward an Integrated Product Quality System 147
7.1 Introduction 147
7.2 The Quality Movement 147
   7.2.1 Quality Control Philosophies 148
      7.2.1.1 Deming Philosophy 148
      7.2.1.2 Crosby Philosophy 149
      7.2.1.3 Juran Philosophy 149
   7.2.2 A Modern View of Quality 149
7.3 A QRW (Quality, Reliability, and Warranty) Paradigm 150
7.4 Some Concluding Remarks 152
References 152

Appendix A Notations and Acronyms 155

Appendix B Selected Tables 159

Appendix C Counting Distributions 167
C.1 Binomial Counting Distribution 167
C.2 Poisson Counting Distribution 168
C.3 Renewal Function for Erlang (λ,2) Distributed Times 170

Appendix D Solutions to Exercises 173

Index 191

Dedication

This book is dedicated to my wife Susan Thomas.

1
Product Quality and Warranty

1.1 Introduction

Manufacturing continues to be the major source for creating wealth and promoting economic growth and development throughout the world. In the U.S. the proportion of jobs in the manufacturing sector has declined for many decades, but the relative impact of job creation in non-manufacturing areas that was generated from manufacturing has increased significantly. This increasing workforce multiplier is due to new and advanced technologies. Global competition has never been greater for introducing and producing new products at reasonable costs that can meet the high expectations of today's well-informed consumers. Successful producers must not only be efficient and cost effective in producing products, but these products must also meet high customer expectations in terms of their usefulness and overall appeal as well as having high performance. Today the public is well-informed of the availability of products and emerging technologies that will impact the markets and people rightfully demand high quality. Essentially every major company is engaged in some form of total quality management program that is aimed at organizing systematic changes in processes that will ensure continuous improvement. To be effective any such program must receive the support and participation of the employees throughout the organization and they need to be knowledgeable of all aspects of their product quality and the factors that can influence it throughout design and development, production, delivery, and customer use. The modern interpretation of quality is broad and incorporates all of the features and characteristics that influence a product's value relative to both consumers and producers of the products. To properly account for the total balance of weight or importance that should be given in making economic decisions, which pertain to product quality, it is necessary to consider a systematic view that incorporates all of the invested interests in achieving customer acceptance and economic feasibility for the manufacturer.

This book provides methods for analyzing and evaluating product quality issues important in developing and producing manufactured products.

These methods deal specifically with warranty information based on the concept that warranty feedback is the key to understanding overall quality. The primary emphasis will be on manufactured products, though the methods can be applied to services as well. The over-arching philosophy for product quality, reliability, and warranty is as follows:

1. Quality is a vector of attributes that relate to the way a product is designed, developed, produced, and accepted by customers.
2. Reliability is a vital element of quality that impacts all other elements.
3. Warranty feedback provides a relative measure of product performance.

A modern interpretation and complete description of product warranty requires multiple characteristics one of which is reliability. These characteristics and their relative importance can vary as they are viewed by customers, manufacturers or producers, and independent agents such as dealers who represent both customers and producers (Garvin, 1987). Since reliability relates to product design and the processes used in production, it influences essentially all other elements of quality and, therefore, is a natural focus for seeking quality improvement. Warranties provide consumers with relief from the threat of quality problems and serve as a marketing tool for manufacturers. Moreover, they provide manufacturers with feedback information on the quality performance of their products. Therefore, it is appropriate to consider quality, reliability, and warranty as a threesome in developing technologies for quality improvement.

1.2 Background and Scope

The word quality has evolved over time and for products it casts a quite different meaning today than it has in the past. Prior to 1900, most products were produced by individuals and families. Quality, as an overall level of satisfaction, was maintained by the self-imposed standards of individual owners who were motivated largely by their pride in workmanship and service. The industrial revolution in the early 1900s brought on the division and specialization of labor which decentralized the personal sense of responsibility for producing good parts and products and necessitated the need for more standards and control methods. During the early years of mass production when the number of manufacturers who produced consumer products was quite small, high volume production typically out weighted testing and inspection options in cost trade-off decisions for ensuring that standards were met since competition was generally not a concern. In spite of the ongoing evolution of quality control methods (see Mitra, 1998), quality was considered almost exclusively as the degree to which production and engineering specifications were achieved. This interpretation held until early 1970 when international competition and new technologies forced a major change in management philosophies, shifting attitudes and consciousness that ultimately altered the meaning of quality throughout the world. By the 1980s, producers were more sensitive to customer needs and customers were becoming more informed on quality and costs. Manufacturers of consumer products started implementing continuous quality improvement programs that would later become standard industry practice.

From an historical perspective reliability emerged independently from quality management and control. Originating from concerns for public safety, fatigue studies of structures and life testing of materials started to appear in the literature in 1929. That same year, Professor Walodie Weibull of the Royal Institute of Technology in Sweden proposed what is now known as the classic Weibull probability distribution for representing the breaking strength of materials, as well as other failure characteristics. Reliability theory as we know it today, emerged in the early 1940s largely from support requirements for World War II. The Department of Defense (DOD) discovered shortly at the onset of the war that over 50% of the airborne electronics equipment that was stored in warehouses as war reserves was found defective or incapable of meeting Air Force and Navy mission requirements. In 1952, DOD established the Advisory Group on Reliability of Electronics Equipment (AGREE) to consolidate the needs for all of the military services to establish measures for increasing the reliability of equipment and reducing maintenance actions. This ultimately led to the first military standard for reliability, MIL-STD-785, published in 1965 that has since been expanded to cover the integration of reliability with design, development, and production functions throughout most industries. These standards are aimed at identifying reliability problems during the embryonic stages of system and product developments.

Today, quality has a much broader interpretation, relating to the way products are designed, produced, and accepted by consumers. Therefore several attributes, one being reliability, are required to completely describe product quality. The relative significance of these attributes will vary among different products, marketing conditions and relative competition. However, reliability is always a key dimension and it significantly impacts customer attitude and acceptance of products. In nearly all cases, quality problems lead to more failures and dissatisfied customers, which result in higher warranty cost to the producer. Likewise, products of high quality have less warranty expenses.

A warranty is a commitment by a manufacturer or producer to provide quality products or services for some specified period of time. An historical account of the evolution of warranties is given by Loomba (1995). The concept of a warranty dates far back in time, but it is only during the past 60 years or so that effective programs have been implemented as we know them today. Today, virtually every consumer product producer offers some type of warranty, either explicitly provided by a formal program or implied through consumer protection legislation (Brennan, 1994). During the warranty period the manufacturer assumes some portion of the expenses that arise from the repair or replacement of defective products. This provides the customer with some relief from the financial risk and inconvenience of failures that might occur during the warranty period. When the quality of the product is high then the cost for the warranty will be relatively low. Similarly, when a low-quality product lot is released, the manufacturer will eventually realize increased costs associated with the warranty. Moreover, if a manufacturer continues to produce items of low quality over time then the market position will also decline and the associated warranty costs could become insurmountable. Indeed, a minimum cost warranty program is an essential element for overall cost control in manufacturing.

Warranties provide consumers with some security against poor quality and are often used by manufacturers as a value added feature to promote a product. This value depends on the quality and expectations by consumers. The warranty policy reflects the manufacturer's view toward the quality of the product and the strategy for marketing it relative to the competition. To do this, of course, it is necessary that the failure characteristics and associated repair and replacement costs be well known in advance. If the manufacturer has a good record of producing high quality products then there is normally less need to offer a lengthy warranty period, unless there is an attempt to penetrate the market. When consumers perceive a product as being of high quality they will perceive the personal risk to be less and therefore will demand less warranty protection. Both the manufacturer and consumer will face some risk due to the randomness and uncertainty with product failures and costs.

Quality is a state of acceptance of how well a product or service is received and used by customers. While several factors can influence acceptance, a very important one is reliability which expresses the level of dependence given to an item that functions properly when in use. Reliability expresses the amount of failure-free time that can be expected from a product. Warranties provide customers with a guarantee of suitable service or functioning for a prescribed period of time. This threesome: quality, reliability, and warranty are fundamental in understanding and predicting failures and product performance over time.

For this book, Chapter 2 begins with some preliminary results for dealing with product failures. Common distributions and methods for analyzing failure times are reviewed, including probability transforms and methods for fitting failure time distributions. There are several characteristics and features that relate to the quality of a given product, all of which ultimately impact on the associated warranty cost. In Chapter 3, we introduce a modern view of quality defined in terms of a set of multidimensional attributes that prescribes the value of an item in terms of the manner in which it is developed, produced, and used by customers. Reliability is one of the most important dimensions of quality. Reliability and quality concepts are presented as they pertain to assessing overall quality. In Chapter 4, warranty is introduced as a relative measure of total quality and methods are presented for using warranty information as feedback for diagnosing and targeting areas for quality improvement. Cost models are developed for the basic types of warranties that are applied to consumer products. Methods for determining optimal warranty periods and determining the amount of reserve funds that should be set aside to cover ensuing warranty expenditures for current sales are presented in this chapter as well. In Chapter 5, warranty is introduced as a relative measure of total quality and methods are presented for using warranty information as feedback for diagnosing and targeting areas for quality improvement. Chapter 6 describes an approach for establishing product quality improvement targets, which includes a decision analysis framework for structuring improvement options, identifying failure modes and criticality analysis, and methods for allocating lower subsystem or element improvement goals. Chapter 7 concludes by providing summary remarks on analyzing and examining product warranties through a quality, reliability, and warranty (QRW) paradigm for understanding and accounting for quality performance. Current and future technologies provide opportunities for sensing and making fast online, real time decisions. This is the path to achieving ultra high reliability goals.

References

Brennan, J.R., 1994, *Warranties: Planning, Analysis, and Implementation*, McGraw-Hill, New York.

Garvin, D.A., 1987, Competing on the eight dimensions of quality, *Harv. Busn. Rev.*, 65, 6: 107-109.

Loomba, A.P.S., 1995, *Product Warranty Handbook*, W.R. Blischke and D.N.P. Murthy, Eds., Marcel Dekker, New York, Ch. 2.

Mitra, A., 1998, *Fundamentals of Quality Control and Improvement*, 2nd ed., Prentice-Hall, Upper Saddle River, NJ, Ch. 2.

2
Preliminary Results for Analyzing Product Failures

Failures occur due to problems in the way a product is developed, produced, and distributed to customers. If a product is well designed and manufactured then it should last or provide effective usage longer than one produced at a lower quality level with an inferior design and poor selection of materials. So a product that has high durability will implicitly also have high conformance quality and reliability. Generally, failures that occur early during their usage are conformance related due to problems in their production, assembly, and distribution. Poor workmanship, impurities, and lack of homogeneity in the materials used cause these problems as well as process-related problems, such as improper sequencing and materials handling. Failures that occur later in the usage cycle are most often due to reliability problems that relate to the design. In all cases, it is very important that the failure characteristics are known before the product is placed on the market. In this chapter, the basic results for predicting and analyzing failures are presented.

2.1 Failure Time Distributions

Let X be a nonnegative, random variable representing the time to failure for a product component or system. We will assume that X has a continuous distribution over the range $(0, \infty)$ with probability density function (pdf) defined by:

$$f(x) = P\{x < X < x + dx\}$$

and such that $f(x) \geq 0$ and

$$\int_0^\infty f(u)du = 1$$

It follows that the probability that a component has failed by time X is given by the cumulative distribution function (cdf)

$$F_X(x) = P\{X \le x\} = \int_0^x f(u)du \qquad (2.1)$$

$F_X(.)$ is nondecreasing in x, $F_X(0) = 0$ since X is nonnegative and $F_X(\infty) = 1$. For cases where $F_X(x)$ is differentiable, as will apply here, it follows that

$$f(x) = \frac{d}{dx} F_X(x)$$

The kth moment of X is defined as

$$E[X^k] = \int_0^\infty x^k f(x)dx \qquad (2.2)$$

from which the mean $\mu = E[X]$ and the variance is given by

$$Var[X] = E[X^2] - E^2[X] \qquad (2.3)$$

2.1.1 Some Common Continuous Distributions

Exponential: Let X be a nonnegative random variable with pdf,

$$f(x) = \lambda e^{-\lambda x}, \lambda > 0, x \ge 0 \qquad (2.4)$$

From (2.2), we have the moments

$$E[X] = \int_0^\infty \lambda x e^{-\lambda x} dx = \frac{1}{\lambda}$$

$$E[X^2] = \int_0^\infty \lambda x^2 e^{-\lambda x} dx = \frac{2}{\lambda^2}$$

and it follows from (2.3) that the variance is

$$Var[X] = \frac{1}{\lambda^2}$$

The exponential distribution plays a major role in reliability since it represents the useful life period for a product or system, and it corresponds to purely random or *chance* events over time.

Preliminary Results for Analyzing Product Failures

Gamma: The pdf for a nonnegative random variable X distributed gamma is given by

$$f(x) = \frac{\lambda(\lambda x)^{r-1} e^{-\lambda x}}{\Gamma(r)}, \quad \lambda > 0, r > 0, x \geq 0 \tag{2.5}$$

where

$$\Gamma(r) = \int_0^\infty x^{r-1} e^{-x} dx \tag{2.6}$$

is the gamma function and, for the special case of r being a positive integer, $\Gamma(r) = (r-1)!$ This distribution is called the Erlang and its cdf can be written

$$F_X(x) = 1 - P\{X > x\} = 1 - \int_x^\infty \frac{\lambda(\lambda w)^{r-1}}{(r-1)!} e^{-\lambda w} dw \tag{2.7}$$

and letting $\lambda w = y$ and, hence, $dw = dy/\lambda$, we have

$$f(x) = 1 - \int_{\lambda x}^\infty \frac{\lambda y^{r-1}}{(r-1)!} e^{-y} dy/\lambda = 1 - \int_{\lambda x}^\infty \frac{y^{r-1}}{(r-1)!} e^{-y} dy$$

and the integral on the right-hand side is of the form

$$I = \int_{\lambda x}^\infty \frac{y^r e^{-y}}{r!} dy$$

where r is integer-valued and $(\lambda x) > 0$, therefore

$$r!I = \int_{\lambda x}^\infty y^r e^{-y} dy$$

By successively integrating by parts, it follows that

$$r!I = e^{-\lambda x}[(\lambda x)^r + r(\lambda x)^{r-1} + r(r-1)(\lambda x)^{r-2} + \cdots + r!]$$

Therefore,

$$I = \sum_{k=0}^r \frac{(\lambda x)^k e^{-\lambda x}}{k!}$$

and it follows that

$$F_X(x) = 1 - \sum_{k=0}^{r-1} \frac{(\lambda x)^k e^{-\lambda x}}{k!} \tag{2.8}$$

The mean and variance are

$$E[X] = \frac{r}{\lambda}, \quad Var[X] = \frac{r}{\lambda^2}$$

Weibull: For a nonnegative random variable X distributed as a two parameter Weibull, the pdf is

$$f(x) = \left(\frac{\beta}{\theta}\right)\left(\frac{x}{\theta}\right)^{\beta-1} e^{-(x/\theta)^\beta}, \quad \beta > 0, \ \theta > 0, \ x \geq 0 \tag{2.9}$$

Letting $u = (x/\theta)^\beta$ and, hence, $du = (\beta/\theta)(x/\theta)^{\beta-1} dx$ in (2.9), it follows that the cdf is given by

$$F_X(x) = 1 - e^{-(x/\theta)^\beta} \tag{2.10}$$

To get the mean and variance, we first determine the kth moment of X as

$$E[X^k] = \int_0^\infty x^k \left(\frac{\beta}{\theta}\right)\left(\frac{x}{\theta}\right)^{\beta-1} e^{-(x/\theta)^\beta} dx$$

Again, using the transformation $u = (x/\theta)^\beta$, it follows that

$$E[X^k] = \theta^k \int_0^\infty u^{k/\beta} e^{-u} du$$

and the integral is the gamma function, therefore

$$E[X^k] = \theta^k \Gamma\left(1 + \frac{k}{\beta}\right) \tag{2.11}$$

The mean and second moments are then

$$E[X] = \theta \Gamma\left(1 + \frac{1}{\beta}\right) \tag{2.12}$$

and

$$E[X^2] = \theta^2 \Gamma\left(1 + \frac{2}{\beta}\right) \qquad (2.13)$$

therefore, the variance is

$$Var[X] = \theta^2 \left[\Gamma\left(1 + \frac{2}{\beta}\right) - \Gamma^2\left(1 + \frac{1}{\beta}\right)\right] \qquad (2.14)$$

The Weibull distribution has wide applications in reliability and it can have a variety of shapes. Here we have the two-parameter Weibull, but it can have additional parameters as well.

Normal: The normal distribution has the pdf

$$f(x) = \frac{1}{\sigma\sqrt{2\pi}} e^{-\frac{1}{2}\left(\frac{x-\mu}{\sigma}\right)^2}, \quad -\infty < x < \infty \qquad (2.15)$$

with parameters $\mu = E[X]$ and $\sigma^2 = Var[X]$. For computational purposes, we translate this to a standard normal by the transformation

$$Z = \frac{X - \mu}{\sigma}$$

which is normally distributed with $\mu = 0$ and $\sigma^2 = 1$. Values for

$$\Phi(z) = \int_{-\infty}^{z} e^{-u^2/2} du \qquad (2.16)$$

are given in standard normal tables.

Though most of the applications of interest in this book deal with nonnegative random variables, the normal distribution is still quite important because the accumulation of a large number of occurring events will converge to the normal. It is also often used in approximating other distributions.

2.1.2 Laplace Transforms

Laplace Transforms (LT) are useful in analyzing stochastic systems, which are described by independent random variables. This is particularly true where it is necessary to sum random variables as in counting events, such as failures, demands, or transactions. The method and properties of the transforms are summarized below. For a more thorough treatment of the subject, the reader is referred to Muth (1977).

Definition: Let f(x) be the pdf of a nonnegative continuous random variable X. The Laplace transform of f(x) is given by

$$\bar{f}(s) = E[e^{-sX}] = \int_0^\infty e^{-sx} f(x)dx \qquad (2.17)$$

Properties — Note that $\bar{f}(0) = 1$ and $0 < \bar{f}(s) < 1$, for all $s \geq 0$ and, if $\bar{f}(s)$ exists, then it will be unique in that there is no more than one f(x) associated with it. The limiting values can be evaluated through the equivalent relationships:

$$\lim_{s \to \infty} s\bar{f}(s) = \lim_{x \to 0} f(x) \qquad (2.18)$$

and

$$\lim_{s \to -\infty} s\bar{f}(s) = \lim_{x \to \infty} f(x) \qquad (2.19)$$

that are sometimes called the initial and final value theorems.

1. Moments

By expanding e^{-sX} in (2.17), we get

$$E[e^{-sX}] = \int_0^\infty \left[1 - sx + \frac{(sx)^2}{2!} - \frac{(sx)^3}{3!} + \cdots \right] f(x)dx$$

$$= \int_0^\infty \sum_{k=0}^\infty \frac{(-sx)^k}{k!} f(x)dx = \sum_{k=0}^\infty \int_0^\infty \frac{(-sx)^k}{k!} f(x)dx$$

therefore,

$$\bar{f}(s) = \sum_{k=0}^\infty \frac{(-s)^k}{k!} E[X^k] \qquad (2.20)$$

Now, by differentiating with respect to s and evaluating the derivatives at $s \to 0$, the moments can be computed from

$$E[X^m] = (-1)^m \frac{d^m}{ds^m} \bar{f}(s) \Big|_{s=0} \qquad (2.21)$$

Example 2.1

Let X be distributed exponential with pdf

$$f(x) = \lambda e^{-\lambda x}, \quad \lambda > 0, \ x \geq 0$$

The Laplace Transform of f(x) is then

$$\bar{f}(s) = \int_0^\infty e^{-sx} \lambda e^{-\lambda x} dx = \lambda \int_0^\infty e^{-(\lambda+s)x} dx$$

$$= \lambda \left. \frac{e^{-(\lambda+s)x}}{-(\lambda+s)} \right|_0^\infty = \frac{\lambda}{\lambda+s}$$

Applying the result of (2.21), we derive the following first and second moments

$$E[X] = (-1) \left. \frac{d}{ds}\left(\frac{\lambda}{\lambda+s}\right) \right|_{s \to 0} = (-1) \left. \frac{-\lambda}{(\lambda+s)^2} \right|_{s \to 0} = \frac{1}{\lambda}$$

$$E[X^2] = (-1)^2 \left. \frac{d^2}{ds^2}\left(\frac{\lambda}{\lambda+s}\right) \right|_{s \to 0} = \left. \frac{2\lambda}{(\lambda+s)^3} \right|_{s \to 0} = \frac{2}{\lambda^2}$$

2. Linearity

Given pdf's $f_1(x)$ and $f_2(x)$ having respective Laplace Transforms $\bar{f}_1(s)$ and $\bar{f}_2(s)$ and a constant $0 < \alpha < 1$, the Laplace Transform of the pdf

$$f(x) = \alpha f_1(x) + (1-\alpha) f_2(x)$$

is given by

$$\bar{f}(s) = \alpha \bar{f}_1(s) + (1-\alpha) \bar{f}_2(s)$$

3. Derivatives and Integrals

Let $\bar{f}(s)$ be the Laplace Transform of a function f(x) that is appropriately differentiable or integrable, then it follows that

$$LT\left\{\frac{d^n}{dx^n} f(x)\right\} = s^n \bar{f}(s) - s^{n-1} \left. \frac{d}{dx} f(x) \right|_{s=0} + s^{n-2} \left. \frac{d^2}{dx^2} f(x) \right|_{s=0} + \cdots + \left. \frac{d^{n-1}}{dx^{n-1}} f(x) \right|_{s=0}$$

$$LT\left\{\int_0^x f(u) du\right\} = \frac{1}{s} \bar{f}(s)$$

4. Sums of Continuous Random Variables

Let X_1 and X_2 be nonnegative independent random variables with pdf's $f_1(x)$ and $f_1(x)$, then the pdf of the random variable $Y = X_1 + X_2$ is given by

$$g(y) = \int_0^y f_1(x_1) f_2(y - x_1) dx_1 \tag{2.22}$$

called the convolution of the pdf's and sometimes symbolized by $g = f_1 * f_2$. It follows that the Laplace Transform of $g(y)$ given in (2.22) is

$$\overline{g}(s) = \overline{f_1}(s) \overline{f_2}(s)$$

where $\overline{f_1}(s)$ and $\overline{f_2}(s)$ are the transforms for $f_1(x)$ and $f_2(x)$. Moreover, this result can be generalized to the sum of n nonnegative random variables for which

$$\overline{g}(s) = \prod_{i=1}^{n} \overline{f_i}(s) \tag{2.23}$$

is the Laplace Transform of $g = f_1^* \cdots {}^* f_n$.

Example 2.2

Consider a machine that breaks down on occasion due to failures that occur at random with a mean time between failures of $1/\lambda$. We want to find the distribution of the time until the rth failure. Therefore, letting $X_1, \ldots, X_r$ be independent and identically distributed exponential random variables representing the times between machine failures, we want to find the distribution of

$$Y = \sum_{i=1}^{r} X_i$$

We could derive the pdf for Y by recursively applying the convolution integral of (2.22). However, it is simpler to use transforms. Since each X_i is distributed exponentially, we found in Example 2.1 that

$$\overline{f}(s) = \frac{\lambda}{\lambda + s}$$

The transform of the pdf for Y is then determined by applying (2.23)

$$\overline{g}(s) = \prod_{i=1}^{r} \left(\frac{\lambda}{\lambda + s} \right) = \left(\frac{\lambda}{\lambda + s} \right)^n$$

which is the transform of the gamma pdf

$$g(x) = \frac{\lambda(\lambda x)^{r-1} e^{-\lambda x}}{(r-1)!}$$

It is often the case that the analysis questions of interest can be obtained directly through the properties of the transform (i.e., in the transform domain). Methods of inversion that include tables and approximations are available for those situations which require a transformation back into the original domain.

2.1.3 Some Common Discrete Distributions

Let X be a discrete random variable that takes on integer values 0,1,2,... and has probabilities assigned in accordance with a discrete probability function (dpf), also called a probability mass

$$p_X(x) = P\{X = x\}, \quad x = 0, 1, 2, \ldots$$

and such that $p_X(x) \geq 0$ and

$$\sum_x p_X(x) = 1$$

It follows that the cdf for X is given by

$$F_X(x) = P\{X \leq x\} = \sum_{j=0}^{x} p_X(j)$$

$F_X(x)$ is nondecreasing in x, $F_X(0) = 0$ for X nonzero, and $F_X(\infty) = 1$. Moreover, it also follows that

$$p_X(x) = F_X(x) - F_X(x-1) \tag{2.24}$$

and the kth moment of X is given by

$$E[X^k] = \sum_{\chi=0}^{\infty} x^k p_X(\chi) \tag{2.25}$$

Geometric: Let X represent the number of cycles, or trials, before the first defect is found in a sequence of Bernoulli trials where the probability of a defect on any given trial is a constant $0 < p < 1$. The dpf for X is then

$$p_X(x) = pq^{x-1}, \quad x = 1, 2, \ldots \tag{2.26}$$

where q = 1–p. To determine the mean, from (2.25) we have

$$E[X] = \sum_{i=1}^{\infty} xpq^{x-1} = p\left[1 + 2q + 3q^2 + 4q^3 + \ldots\right]$$

$$= p\frac{d}{dq}\left[q + q^2 + q^3 + \ldots\right] = p\frac{d}{dq}\left(\frac{q}{1-q}\right)$$

from which it follows that

$$E[X] = \frac{1}{p}$$

Following a similar procedure for computing the second moment, we can derive the variance

$$Var[X] = \frac{q}{p^2}$$

The geometric distribution is the discrete analog to the exponential and it possesses similar properties relative to failure characteristics and reliability.

Negative Binomial: Consider a sequence of independent Bernoulli trials and let X represent the number of trials before the rth defect occurs.

$$p_X(x) = \binom{x-1}{r-1} p^r q^{x-r}, \quad x = r, r+1, r+2, \ldots \qquad (2.27)$$

This is also called the Pascal distribution and it is the discrete analog to the continuous Erlang distribution. The mean and variance are

$$E[X] = \frac{r}{p}, \quad Var[X] = \frac{rq}{p^2}$$

2.1.4 Probability Generating Functions

Laplace Transforms are useful in analyzing stochastic systems described by continuous random variables. For those systems containing elements involving discrete random variables, probability generating functions (pgf) are commonly applied. The pgf, also called the geometric transform, is one of several versions of power transformation that are used in probabilistic models of operational systems. The method and properties are summarized as follows:

Preliminary Results for Analyzing Product Failures

Definition: Let $p_X(x)$ be the dpf for a nonnegative random variable X. The pgf is given by

$$P(z) = E[z^X] = \sum_{x=0}^{\infty} z^x p_X(x) \qquad (2.28)$$

Properties of pgf's — We note that P(z) is continuous in z and converges over $|z| \leq 1$, $P(0) = p_X(0)$, and $P(1) = 1$. As with the Laplace Transform, for a given dpf there will be one and only one associated pgf and vice versa, provided it exists.

1. Probabilities

For a given pgf, the probabilities can be determined by computing the derivatives and evaluating them at $z = 0$. Let

$$P^{(n)}(z) = \frac{d^n}{dz^n} P(z)$$

therefore

$$P^{(n)}(0) = \frac{d^n}{dz^n} P(z) \bigg|_{z=0}$$

$$P^{(1)}(0) = \sum_{n=1}^{\infty} n z^{n-1} \bigg|_{z=0} = \left\{ p_X(1) + 2z p_X(2) + 3z^2 p_X(3) + \ldots \right\}_{z=0} = p_X(1)$$

$$P^{(2)}(0) = \sum_{n=2}^{\infty} n(n-1) z^{n-2} \bigg|_{z=0} = \left\{ 2 p_X(2) + 6z p_X(3) + 12z^2 p_X(4) + \ldots \right\}_{z=0} = 2 p_X(2)$$

$$P^{(3)}(0) = \sum_{n=3}^{\infty} n(n-1)(n-2) z^{n-3} \bigg|_{z=0} = \left\{ 6 p_X(3) + 24z p_X(4) + 60z^2 p_X(5) + \ldots \right\}_{z=0}$$

$$= 6 p_X(3)$$

...

$$P^{(m)}(0) = m! \, p_X(m), \quad m = 0, 1, 2, \ldots$$

or

$$p_X(x) = \frac{1}{x!} P^{(x)}(0) \qquad (2.29)$$

2. Factorial Moments

The pgf can also be used to generate the moments of a distribution. It is easily shown that the mth factorial moment is given by

$$E[X^{[m]}] = P^{(m)}(1) = \left. \frac{d^m}{dz^m} P(z) \right|_{z=1} \qquad (2.30)$$

where

$$X^{[m]} = X(X-1)(X-2)\cdots(X-m+1)$$

Example 2.3

Let X be distributed uniformly over the integers (1, 2, 3, 4),

$$p_X(x) = \begin{cases} 1/4, & x = 1,2,3,4 \\ 0, & \text{otherwise} \end{cases}$$

The pgf is then

$$P(z) = \frac{1}{4}(z + z^2 + z^3 + z^4)$$

which has derivatives

$$P^{(1)}(z) = \frac{1}{4}(1 + 2z + 3z^2 + 4z^3 + \cdots)$$

$$P^{(2)}(z) = \frac{1}{4}(2 + 6z + 12z^2)$$

$$P^{(3)}(z) = \frac{1}{4}(6 + 24z)$$

$$P^{(4)}(z) = \frac{1}{4}(24)$$

So, to find $P\{X \leq 2\}$ from (2.29)

$$p_X(1) = \frac{1}{1!} P^{(1)}(0) = 1 \cdot \frac{1}{4} = \frac{1}{4}$$

$$p_X(2) = \frac{1}{2!} P^{(2)}(0) = \frac{1}{2} \cdot \frac{2}{4} = \frac{1}{4}$$

and

$$P\{X \leq 2\} = \frac{1}{4} + \frac{1}{4} = \frac{1}{2}$$

To find the mean and variance of X, from (2.30) we have

$$E[X] = E[X^{[1]}] = P^{(1)}(1) = \frac{5}{2}$$

$$E[X^{[2]}] = E[X(X-1)] = P^{(2)}(1) = 5$$

and, thus,

$$Var[X] = E[X^2] - E^2[X]$$
$$= E[X^{[2]}] + E[X] - E^2[X]$$
$$= 5 + \frac{5}{2} - \left(\frac{5}{2}\right)^2 = \frac{5}{4}$$

3. Linearity

Given dpf's $p_1(x)$ and $p_2(x)$ having respective pgf's $P_1(z)$ and $P_2(z)$ and a constant $0 < \alpha < 1$, the pgf of the distribution

$$p(x) = \alpha p_1(x) + (1-\alpha) p_2(x)$$

is given by

$$P(z) = \alpha P_1(z) + (1-\alpha) P_2(z)$$

4. Sums of Discrete Random Variables

Let X_1 and X_2 be nonnegative and independent discrete random variables with dpf's $p_1(x)$ and $p_2(x)$, then the dpf of the random variable $Y = X_1 + X_2$ is the convolution $g = p_1 * p_2$ given by

$$g(y) = \sum_{x=0}^{y} p_1(x) p_2(y-x) \qquad (2.31)$$

It follows that the pgf for $g(y)$ is

$$G(z) = P_1(z) P_2(z) \qquad (2.32)$$

It also follows that for $Y = \sum_{i=1}^{n} X_i$ independent discrete random variables $g = p_1 * \ldots * p_n$ has the pgf given by

$$G(z) = \prod_{i=1}^{n} P_i(z) \qquad (2.33)$$

Example 2.4
Consider the independent random variables X_1 and X_2 distributed geometric with dpf's

$$p_1(x) = \begin{cases} 0.2(0.8)^{x-1}, & x = 1, 2, \ldots \\ 0, & \text{otherwise} \end{cases}$$

and

$$p_2(x) = \begin{cases} 0.3(0.7)^{x-1}, & x = 1, 2, \ldots \\ 0, & \text{otherwise} \end{cases}$$

The pgf can be found in tables of probability transformations or computed directly using (2.28). For the geometric, we have

$$P(z) = \sum_{x=1}^{\infty} z^x p q^{x-1} = pz \sum_{x=1}^{\infty} (qz)^{x-1}$$

or

$$P(z) = \frac{pz}{1 - qz} \qquad (2.34)$$

Therefore, the pgfs for $p_1(x)$ and $p_2(x)$ are

$$P_1(x) = \frac{.2z}{1 - .8z}, \quad P_2(z) = \frac{.3z}{1 - .7z}$$

and, from (2.33) for $Y = X_1 + X_2$, the pgf is

$$G(z) = \left(\frac{.2z}{1 - .8z}\right)\left(\frac{.3z}{1 - .7z}\right)$$

or

$$G(z) = \frac{.06z^2}{(1 - .8z)(1 - .7z)} \qquad (2.35)$$

5. Inversion Procedure

Inversion is not always necessary since the probabilities and moments can be derived from the pgf using (2.29) and (2.30). Whenever it is necessary to invert a pgf, the first step is to algebraically convert the pgf into a sum of terms that have known transforms using partial fractions. We will demonstrate this method in Example 2.4.

Preliminary Results for Analyzing Product Failures

To invert the pgf in (2.35), we note that the numerator is of the same degree as the denominator. Therefore, to get G(z) in the proper format we perform division to obtain

$$G(z) = \frac{3}{28} + \frac{(9z/56) - (3/28)}{(1-.8z)(1-.7z)}$$

The constant term can be disregarded, so we are to find A and B such that

$$\frac{(9z/56) - (3/28)}{(1-.8z)(1-.7z)} = \frac{A}{(1-.8z)} + \frac{B}{(1-.7z)}$$

Multiplying through by (1−.8z) and then (1−.7z), we obtain the equations

$$\frac{(9z/56) - (3/28)}{(1-.7z)} = A + B\left(\frac{1-.8z}{1-.7z}\right) \tag{2.36a}$$

$$\frac{(9z/56) - (3/28)}{(1-.8z)} = A\left(\frac{1-.7z}{1-.8z}\right) + B \tag{2.36b}$$

Letting $z \to 1/.8$ in (2.36a) and $z \to 1/.7$ in (2.36b), we find that A = 3/4 and B = −6/7, therefore

$$G(z) = \frac{3}{4(1-.8z)} - \frac{6}{7(1-.7z)}$$

which can be easily inverted from tables (e.g., see Muth, 1977) to obtain

$$g(y) = 0.6(0.8)^{x-1} - 0.6(0.7)^{x-1}, \quad x = 2,3,\ldots \tag{2.37}$$

2.2 Fitting Distributions

In analyzing failures, one typically has to establish the choice of failure time distribution through some selection process using data. This is generally accomplished by postulating a particular choice of distribution and then verifying the assumption through analysis of the data, using as much statistical strength as you can gain through the given data and acceptable risk level. There are two distinct situations to consider in analysis. The first is for parametric methods where one can assume a hypothetical distribution and then test for goodness of fit using standard test statistics. The other situation is where one is unable to rationally assume a particular distribution and, therefore, must resort to *distribution-free* methods, which are also called *nonparametric* methods for fitting data. Here we will consider the basic goodness of fit test based on parametric methods. Ebeling (2005) discusses nonparametric methods including those for dealing with small sample sizes and censored data.

2.2.1 Goodness of Fit Testing

Let $x_1,\ldots,x_n$ be a random sample from a failure time population represented by the random variable X with distribution F(x;p), where p is the set of parameters for F. Thus, the samples x_i are independent and identically distributed. The procedure is to construct a hypothesis that the data come from some given distribution F_0 against an alternative distribution, say F_1. We then compare the observed frequency distribution $\hat{F}$ from the data with corresponding expected values from the theoretical distribution, as illustrated in Figure 2.1.

To test the hypothesis: $H_0: F = F_0$ vs. $H_1: F = F_1$.
compute test statistic

$$\chi^2 = \sum_{j=1}^{k} \frac{(f_j - e_j)^2}{e_j} \quad (2.38)$$

where

$$e_j = n\left[F_0(j) - F_0(j-1)\right] \geq 5, \; j = 1,\ldots,k; \; F_0(0) = 0 \quad (2.39)$$

is the expected number of events in a cell j from the theoretical distribution and f_j is the number of observations from the sample data in cell j. The expected frequency of events in each of the $k \leq n$ cells must be at least 5.

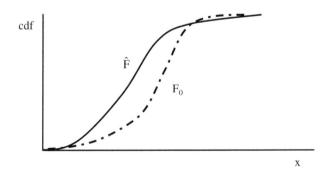

FIGURE 2.1
Representation of fitting distributions.

Preliminary Results for Analyzing Product Failures

The test statistic χ^2 is asymptotically for large n distributed as a chi-square distribution with k-p-1 degrees of freedom. The procedure is then to reject the hypothesis if the computed value of the statistic in (2.38) is greater than the value $\chi^2_{\alpha;k-p-1}$ from the chi-square tables for a level of significance $0 < \alpha < 1$. Otherwise we conclude that there is no reason to not accept F_0 at the α level of significance.

Example 2.5

The following 30 machine failures reported in months of service have been collected for a maintenance facility. Determine if a normal distribution will fit these data at the 5% level of significance.

89	73.1	71.9	62.4	68.6	83.9
68.8	85.4	77.5	64.6	99	58.8
74.3	88.9	72	84.8	75.6	86.9
70	69.3	79.5	85.2	86.3	82.6
89.8	76	75.7	78.4	71.3	75.4

So, F_0 is presumed to be a normal distribution and we estimate the parameters μ and σ^2 by the sample mean $\bar{x} = 77.5$ and variance $s^2 = 84.62$ computed from the data. Dividing the data into cells of 5-month increments we get the frequency count of observations shown in Table 2.1.

The expected number of observations in cell 1 from (2.39) given the theoretical distribution normal is

$$e_1 = 30 F_0(1) = 30\Phi\left(\frac{65 - 77.5}{9.2}\right) = 30(0.0871) = 2.6$$

TABLE 2.1
Goodness of Fit Summary for Monthly Machine Failures

Cell i	Interval	f_j		e_j		$(f_j - e_j)^2 / e_j$
1	x<65	3	6	2.6	6.2	0.01
	65≤x<70	3		3.6		
2	70<x<75	7		5.6		0.35
3	75<x<80	6		6.4		0.025
4	80<x<85	3		5.6		1.21
5	85 < x < 90	7		3.6		
	90 < x < 95	1	8	1.8	6.2	0.52
	x > 95	0		0.85		
Sum						2.115

and for cell 2,

$$e_2 = 30\left[\Phi\left(\frac{70-77.5}{9.2}\right) - \Phi\left(\frac{65-77.5}{9.2}\right)\right]$$

$$= 30\left[\Phi(-0.8152) - \Phi(-1.359)\right] = 3.6$$

and, similarly for e_3 to e_6, noting that the last three cells had to be combined in order for $e_6 \geq 5$. The value of the test statistic is

$$\chi^2 = \sum_{j=1}^{5} \frac{(f_i - e_i)^2}{e_i} = 2.115$$

With k = 5 cells and p = 2 parameters, the degrees of freedom for the decision cut-off criteria is 2 and $\chi^2_{0.05;2} = 5.99$ from the tables, which is significantly greater than χ^2. We, therefore, conclude that there is no reason to reject the hypothesis of the failures coming from a normal distribution.

2.2.2 Probability Plotting

An alternative method for fitting a distribution is to estimate a fit using probability plotting. This graphical technique lacks the statistical strength of other methods such as the chi-square, but it is quick and useful for cases where small sample sizes are necessary for use in establishing a distribution. The procedure is to find an algebraic translation that will make it easy to fit the data using a graphical straight-line relationship.

For a given sample $x_1, x_2, ..., x_n$, the observations are first arranged in rank order $x_{[1]}, x_{[2]}, ..., x_{[n]}$, where $x_{[1]}$ is the smallest value and $x_{[n]}$ the largest. The cumulative frequencies, (j-0.5)/n (also called plotting positions), are then computed for $\hat{F}(x_{[j]})$ from this ordered arrangement. In order to obtain a linear fit on Cartesian coordinates, an appropriate translation of the distribution, $G[\hat{F}(x)]$, must be applied. For the case of an exponential distribution,

$$F(x) = 1 - e^{-\lambda x}$$

and it follows that since

$$G[F(x)] = \ln\left[\frac{1}{1-F(x)}\right] = \lambda x \qquad (2.40)$$

a plot of the points $(x_{[j]}, G[(j-0.5)/n])$ should produce a straight line fit.

For fitting a normal distribution, it follows that since

$$F(x) = \Phi\left(\frac{x-\mu}{\sigma}\right) \qquad (2.41)$$

then, for a given value x_i, there is a corresponding value

$$z_i = \Phi^{-1}\left[F(x_i)\right] = \frac{1}{\sigma}x_i - \frac{\mu}{\sigma} \tag{2.42}$$

Therefore, letting $G(x) = \Phi^{-1}[F(x)]$, we can develop a linear fit if F is normal.

Example 2.5 (continued)

To develop a normal probability plot for the monthly machine failure data, we first order the sample and align the points with their respective plotting positions as given in Table 2.2. Columns 2, 5, and 8 are the z_i values from the inverse normal distribution of (2.42). The resulting normal plot is given in Figure 2.2 and shows a relatively good fit of these data by the line $\hat{z} = 0.1073x - 8.314$. The parameters are then computed by matching the coefficients of (2.42) with the slope $\frac{1}{\sigma} = 0.1073$ and intercept $\frac{\hat{\mu}}{\sigma} = 8.314$ and solving to obtain the estimates $\hat{\sigma} = 9.32$ and $\hat{\mu} = 7.48$. Depending upon the degree of accuracy being sought, this line could be estimated by simply eyeballing the data, or by actually computing the regression line. The result shows a very good fit, which we expect since the data had a very good fit using the chi-square goodness of fit test.

The chi-square goodness of fit test is very common when data are available and of a reasonable sample size. There are other methods, such as the Kolmogorov-Smirnov nonparametric test, and specialized tests for the exponential and Weibull distributions when sample sizes are relatively small. Probability plotting is useful for conducting a preliminary analysis or working with small samples. Special graph paper that is appropriately scaled for distributions, such as the normal, Weibull, and gamma, are available for probability plotting as an option to computing the translation scales for Cartesian coordinates. Further details on these and other methods for fitting distributions are described in Ebeling (2005).

TABLE 2.2

Data Table for a Normal Plot of Machine Failures

i	$\Phi^{-1}\left[\frac{i-0.5}{30}\right]$	x_i	i	$\Phi^{-1}\left[\frac{i-0.5}{30}\right]$	x_i	i	$\Phi^{-1}\left[\frac{i-0.5}{30}\right]$	x_i
1	−2.128	58.8	11	−0.3853	73.1	21	0.47704	83.9
2	−1.6449	62.4	12	−0.2967	74.3	22	0.57297	84.8
3	−1.383	64.6	13	−0.2104	75.4	23	0.67449	85.2
4	−1.1918	68.6	14	−0.1257	75.6	24	0.7835	85.4
5	−1.0364	68.8	15	−0.0418	75.7	25	0.90273	86.3
6	−0.9027	69.3	16	0.04179	76	26	1.03643	86.9
7	−0.7835	70	17	0.12566	77.5	27	1.19182	88.9
8	−0.6745	71.3	18	0.21043	78.4	28	1.38299	89
9	−0.573	71.9	19	0.29674	79.5	29	1.64485	89.8
10	−0.477	72	20	0.38532	82.6	30	2.12804	99

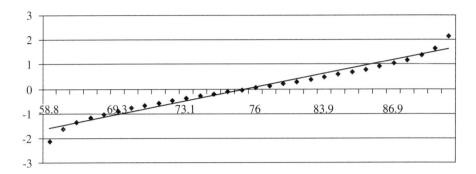

FIGURE 2.2
A normal plot of $\Phi^{-1}[(i-0.5)/30]$ vs. $x_{[i]}$ for machine failures.

2.3 Counting the Number of Failures

Let $T_1, T_2, \ldots$ be a sequence of failure times and $X_n = T_n - T_{n-1}$ represents the times between failures, as shown in Figure 2.3. N(t) for a fixed $t \geq 0$ is a random variable that counts the number of such failures that occurs in an interval (0,t). If the sequence of random variables $X_1, X_2, X_3, \ldots$ are independent and identically distributed F, then the counting process $\{N(t): t \geq 0\}$ is an ordinary renewal process. For the special case of a renewal process with X_1 distributed F_1 that is different from the common distribution for X_2, X_3, then $\{N(t): t \geq 0\}$ is called a delayed renewal process. Delayed renewal processes arise where the first event is quite different from subsequent events. An example might be a new machine that is installed in a facility and the first breakdown occurs earlier than usual due to excessive abuse of the equipment in the initial training of operators. This will be considered as a special case and for our purpose here when we refer to a renewal process it will be an *ordinary* renewal process.

The mean value function for a renewal process is called the *renewal function* and it gives the mean number of renewal events that occur over an interval

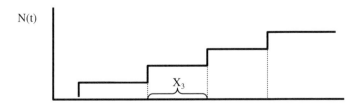

FIGURE 2.3
Representation of a renewal counting process.

of time. To arrive at this function, we note in Figure 2.3 that the time of the nth renewal is given by

$$S_n = \sum_{i=1}^{n} X_i, \quad n \geq 1 \qquad (2.43)$$

and, which is related to N(t) as follows:

$$N(t) \geq n \quad \text{if and only if} \quad S_n \leq t$$

Therefore,

$$P\{N(t) = n\} = P\{N(t) \geq n\} - P\{N(t) \geq n+1\}$$
$$= P\{S_n \leq t\} - P\{S_{n+1} \leq t\}$$

and, since $X_1,\ldots X_n$ are independent with common distribution F, it follows that S_n has the distribution given by $F^{(n)}(t)$, the n-fold convolution of $F_X(x)$ with itself. Therefore,

$$M(t) = E\big[N(t)\big] = \sum_{n=1}^{\infty} F^{(n)}(t) \qquad (2.44)$$

where

$$F^{(n)}(t) = \begin{cases} \displaystyle\sum_{x=0}^{t} F^{(n-1)}(x)F(x-t), & X \text{ discrete} \\ \displaystyle\int_{0}^{t} F^{(n-1)}(x)F(x-t)du, & X \text{ continuous} \end{cases}$$

Applying (2.44) directly for a given failure time distribution $F_X(x)$ can be cumbersome due to the convolutions. Alternatively, M(t) can often be explored through transforms.

2.3.1 Binomial Failure Process

Let the failure times X_i be independent and distributed geometric with dpf from (2.26)

$$p_X(x) = pq^{x-1}, \quad 0 < q = 1-p < 1, x = 1,2,\ldots$$

It follows (see Appendix C.1) that, for this case, the distribution of the number of failures, N in a fixed number of cycles (0,n) is distributed binomial with dpf

$$P(N = x) = \binom{n}{x} p^x q^{n-x}, \quad x = 0,1,\ldots,n \qquad (2.45)$$

with E[N] = np. This is the renewal function for the binomial.

For the case of discrete failure times, the pgf of M(t) is given by

$$M(z) = \sum_{t=0}^{\infty} z^t \sum_{n=1}^{\infty} F^{(n)}(t) = \sum_{n-1}^{\infty} \sum_{t=0}^{\infty} z^t F^{(n)}(t)$$

or

$$M(z) = \sum_{n=1}^{\infty} F^{(n)}(z) \tag{2.46}$$

where $F^{(n)}(z)$ is the pgf of the distribution from the n-fold convolution of $F_X(t)$ and from (2.33)

$$M(z) = \sum_{n=1}^{\infty} \left[F_X(z) \right]^n \tag{2.47}$$

It can be shown that if $G^F(z)$ is the pgf for a cumulative distribution function F, then the pgf for its associated dpf is related by

$$G^f(z) = \left(\frac{1}{1-z} \right) G^F(z) \tag{2.48}$$

Therefore, in (2.46)

$$M(z) = \sum_{n=1}^{\infty} \left(\frac{1}{1-z} \right) \left[P_X(z) \right]^n = \frac{P_X(z)}{(1-z)(1-P_X(z))}, \quad |P_X(z)| < 1 \tag{2.49}$$

where $P_X(z)$ is the pgf for the failure time dpf.

Verifying this through (2.49), we have from (2.34) the pgf for the geometric distribution is

$$P_X(z) = \frac{pz}{1-qz}$$

therefore,

$$M(z) = \left(\frac{1}{1-z} \right) \left[\frac{\dfrac{pz}{1-qz}}{1 - \dfrac{pz}{1-qz}} \right] = \frac{pz}{(1-z)^2} \tag{2.50}$$

which inverts to the renewal function $M(n) = np$.

2.3.2 Poisson Failure Process

Let the failure times X_i be independent and distributed exponentially with probability density function from (2.4)

$$f(x) = \lambda e^{-\lambda x}, \quad \lambda > 0, \ x \geq 0$$

Under these conditions, it follows (see Appendix C.2) that the number of failures occurring in a fixed interval of time (0,t) is distributed Poisson with a rate of λ failures per unit time,

$$p_{N(t)}(t) = \frac{(\lambda t)^n e^{-\lambda t}}{n!}, \quad n = 0, 1, \ldots \qquad (2.51)$$

Therefore, the mean number of failures in (0,t) is known to be the mean of the Poisson, which is easily shown to be $E[N(t)] = \lambda t$.

Verifying this, using the renewal function development in (2.44), since the failure times are continuous, we first take the Laplace transform of M(t)

$$\overline{M}(s) = \int_0^\infty e^{-st} \sum_{n=1}^\infty F^{(n)}(t) dt = \sum_{n=1}^\infty \int_0^\infty e^{-st} F^{(n)}(t) dt$$

or

$$\overline{M}(s) = \sum_{n=1}^\infty \frac{1}{s}\left[\overline{f}(s)\right]^n = \frac{\overline{f}(s)}{s(1-\overline{f}(s))}, \quad |\overline{f}(s)| < 1 \qquad (2.52)$$

because for probability density and distribution functions f(t) and F(t), their respective Laplace transforms are related by

$$\overline{F}(s) = \frac{1}{s}\overline{f}(s) \qquad (2.53)$$

and $F^{(n)}(t)$ is indeed a cumulative distribution function.

Now for the exponential distributed failure times, the Laplace transform of the pdf is

$$\overline{f}(s) = \frac{\lambda}{\lambda + s}$$

and substituting into (2.52), we have

$$\overline{M}(s) = \frac{\dfrac{\lambda}{\lambda+s}}{s\left[1 - \dfrac{\lambda}{\lambda+s}\right]} = \frac{\lambda}{s^2}$$

which has the inversion

$$M(t) = \lambda t$$

Both the binomial and Poisson renewal counting processes are fundamental in product reliability; the binomial for when failure times are measured as discrete counts and the Poisson for continuous time. Ideally, when a product enters the market, its failure characteristics are known as well as the expectations of the customers or users. All of the processes involved in manufacturing and producing the items can be controlled and the only failures that occur are those due to pure chance and not systematic errors from problems in the actual production. Under these conditions, a product enters the market functioning in its useful life.

2.3.3 Approximations to M(t)

For various reasons many products are sold that have relatively short periods of useful life and, thus, their rate of occurrence of failures increases over time. This can be due to an item being introduced into the market before it has been adequately tested and evaluated. It can also be due to a model change in a product that has been produced for several years, but the changes and upgrades cause a change in the failure characteristics of other components. For example, a new model of a vehicle might have an electronic ignition that improves the capability of starting the engine, but causes false alarms in the sensors that display diagnostic problems. One or more elements of a product system can give rise to an early state of increasing failures, thus resulting in M(t) no longer being a simple linear function of the failure rate are the cases for the binomial and Poisson renewal counting processes.

The failure time distributions that are most common in reliability applications are the exponential, Weibull, gamma, and s-normal. With the exception of the exponential, none of these distributions will form an explicit closed-form result for M(t). Results are available, however, for constructing accurate approximations and bounds.

Weibull Distributed Failure Times

Leadbetter (1963) developed an approximation method for the case of X_i distributed Weibull with

$$F(t) = 1 - e^{-(t/\theta)^\beta} = \sum_{r=1}^{\infty} \frac{(-1)^{r-1}}{\Gamma(1+r\beta)} C_r ((t/\theta)^\beta)^r \qquad (2.54)$$

where

$$C_r = \frac{\Gamma(1+r\beta)}{r!}$$

Preliminary Results for Analyzing Product Failures

The renewal function, given by

$$M(t) = \sum_{r=1}^{\infty} \frac{(-1)^{r-1}}{\Gamma(1+r\beta)} A_r ((t/\theta)^\beta)^r \qquad (2.55)$$

with $A_1 = C_1$, $A_2 = C_2 - A_1 C_1$ and $A_j = C_j - \sum_{r=1}^{j-1} A_r C_{j-r}$, $j \geq 2$, is approximated for $t/\theta < 1$ by the truncated series

$$M_L(t) = \sum_{r=1}^{K} \frac{(-1)^{r-1}}{\Gamma(1+r\beta)} A_r ((t/\theta)^\beta)^r \qquad (2.56)$$

To illustrate the method, we consider the following example.

Example 2.6

The times between failures, X_i for a mechanical device, are independent and identically distributed Weibull with shape parameter $\beta = 2$ and scale parameter $\theta = 1$ year. Approximate the mean number of failures during warranty periods of 3 months, 6 months, and 1 year. Apply the approximation of (2.56) with $K = 3$.

Since r is integer valued, $\Gamma(1+2r) = (2r)!$, therefore

$$M_L(t) = \sum_{r=1}^{3} \frac{(-1)^{r-1}}{(2r)!} A_r t^{2r} \qquad (2.57)$$

and

$$C_r = \frac{(2r)!}{r!} = (2r)(2r-1)(2r-2)\ldots(r+1)$$

To approximate $M(t)$, we start by computing the coefficients:

$$C_1 = 2$$
$$C_2 = (4)(3) = 12$$
$$C_3 = (6)(5)(4) = 120$$

Therefore,

$$A_1 = C_1 = 2$$
$$A_2 = C_2 - A_1 C_1 = 12 - (2)(2) = 8$$
$$A_3 = C_3 - [A_1 C_2 + A_2 C_1] = 120 - [(2)(12) + (8)(2)] = 80$$

and substituting into (2.57)

$$M_L(t) = t^2 - \frac{1}{3}t^4 + \frac{1}{9}t^6$$

It follows that for values of t equal to $1/4$, $1/2$, and 1 year, the approximations for $M(1/4) = 0.06123$, $M(1/2) = 0.2309$, and $M(1) = 0.7778$.

Asymptotic M(t)

A very old, but commonly applied, asymptotic form of M(t) by Smith (1954) provides the approximation

$$M_A(t) = \frac{t}{\mu} + \frac{\sigma^2 - \mu^2}{2\mu^2} \qquad (2.58)$$

for the case where the mean μ and variance σ^2 of the time between failures, X_i are finite. This approximation is limited to large values of t.

Applying this result for the Weibull distributed times in Example 2.6 with $\beta = 2$ and $\theta = 1$, the mean and variance of X from (2.12) and (2.14) are

$$\mu = \theta \Gamma\left(1 + \frac{1}{\beta}\right) = \Gamma(3/2) = 0.8862$$

$$\sigma^2 = \theta^2 \left\{ \Gamma\left(1 + \frac{2}{\beta}\right) - \left[\Gamma\left(1 + \frac{1}{\beta}\right)\right]^2 \right\} = \Gamma(3) - \left[\Gamma(3/2)\right]^2 = 1.215$$

Substituting into (2.58), this approximation is

$$M_A(t) = 1.128t + 0.2735$$

For t = 1, this approximation is $M_A(1) = 1.402$, which, as expected, does not compare well with the Leadbetter approximation. $M_A(t)$ gives a good approximation to M(t) when t is very large.

Other methods for approximating M(t) under various conditions include series approximations, bounding techniques, and numerical methods. For details on these methods, see Blischke and Murthy (1994).

2.4 Exercises

1. Show that the Laplace Transform for the gamma probability density function with parameters λ and r, given in (2.5) is

$$\bar{g}(s) = \left(\frac{\lambda}{\lambda + s}\right)^r$$

Verify that the mean and variance are

$$E[X] = r/\lambda, \quad Var[X] = r/\lambda^2$$

2. Show that the Laplace Transform of the probability density function for a normally distributed random variable with mean μ and variance σ^2 is given by

$$\bar{f}(s) = e^{-\mu s + \frac{1}{2}\sigma^2 s^2}$$

3. Show that for n independent and identically distributed normal random variables the sum $Y = X_1 + \cdots + X_n$ is also normally distributed with mean $n\mu$ and variance $n\sigma^2$.

4. Let X_1 and X_2 be independent random variables each distributed uniform over (0,a). Determine the probability density function for $Y = X_1 + X_2$ using the convolution integral of (2.22).

5. A discrete random variable X takes on values $x \in (0, 1, 2, 3)$ according to the discrete probability function

$$p_X(x) = \begin{cases} 1/8, & x = 0 \\ 1/4, & x = 1 \\ 1/2, & x = 2 \\ 1/8, & x = 3 \end{cases}$$

Compute the probability generating function for $p_X(x)$ and apply it in answering the following:

a. Derive $P(X \leq 2)$, $E[X]$, and $Var[X]$.

b. Suppose there are two random variables X_1 and X_2 that are independent and distributed $p_X(x)$. Find $E[X_1 + X_2]$ and $Var[X_1 + X_2]$.

6. Let $X_1, \ldots, X_n$ be independent random variables with discrete probability functions, $p_{X_i}(x)$ and associated probability generating functions, $P_{X_i}(z)$, $i = 1, \ldots, n$. Prove that the probability generating function for $Y = X_1 + \cdots + X_n$ is given by (2.33).

7. The following data represent failure times in 100 hours of operation for a device that is presumed to have exponentially distributed failure times.

10	0.9	23.1	23.2	15.4	41.4	8.6	4.7	60.2	3.2
47.9	8.6	13.7	19.7	2.5	3	19.5	10.3	9.8	88.1
5.8	2.7	3.2	32.8	6	12.9	2.1	2.9	9.7	1.4

Determine if these data can be fitted by an exponential distribution at the 5% level of significance.

8. Let $X_1, X_2, ..., X_r$ be a sequence of independent failure times each identically distributed exponentially with mean $1/\lambda$. Apply the convolution integral of (2.22) to show that the time to the rth failure

$$S_r = \sum_{i=1}^{r} X_i$$

has the probability density function

$$g_{S_r}(y) = \frac{\lambda(\lambda y)^{r-1}}{(r-1)!} e^{-\lambda y}, \quad \lambda > 0, \ r = 1, 2, ...; \ y \geq 0$$

9. Verify the result in Exercise 7 using transforms.
10. The number of failures that occur in (0,t) is distributed Poisson with

$$p_N(n) = \frac{(\lambda x)^n}{n!} e^{-\lambda x}, \quad \lambda > 0, \ n = 0, 1, ...$$

but at each failure time there is a constant probability $0 < \gamma < 1$ that the failure is of Type A. Otherwise, it is a Type B failure and γ is independent of the occurrences. Derive the distribution for the number of Type B failures.

11. The failure time in 100 hours of operation of a device is known to be Weibull distributed with parameters $\beta = 2$ and $\theta = 4$.

    a. Apply the Leadbetter approximation with K = 3 to estimate the mean number of failures in 50 hours and 100 hours of operation.
    b. Compare the results of the above with the asymptotic approximation of (2.58).

References

Blischke, W.R. and D.N.P. Murthy, 1993, *Warranty Cost Analysis*, Marcel Dekker, New York, Ch. 3.

Ebeling, C.E., 2005, *An Introduction to Reliability and Maintainability Engineering*, Chapter 12, Waveland Press, Long Grove, IL.

Leadbetter, M.R., 1963, On series expansion for the renewal moments, *Biometrika*, 50, No. 1/2, 75–80.

Muth, E.J., 1977, *Transform Methods with Applications to Engineering and Operations Research*, Ch. 7, App. 4, Prentice-Hall, Englewood Cliffs, NJ.

Smith, W.L., 1954, Asymptotic renewal theorems, *Proc. Royal Soc.*, Edinburgh, Scotland, 64, : 9–48.

3

Quality and Reliability

3.1 Quality Concepts

The word quality has long been a household word that broadly represents the level of approval people assign to a product or service. While the interpretation and reasons for their opinions vary among different people due to varying roles and needs, the preferences that people derive are based on how a product is made and how it will serve their individual needs. This, of course, is extremely important in marketing products; therefore, it is essential to understand the true meaning of quality and how it is influenced by the way products are produced.

Definition — *Product quality is a state of acceptance of a product or service for the satisfaction customers receive relative to given requirements.*

Product quality is the outcome of an array of factors and attributes that relate to the value that is assigned for the way an item is developed, produced, and used by consumers. Garvin (1987) proposed eight dimensions for defining quality through characteristics relative to the manufacturer and customer, and as a product system. In most cases, the manufacturer is concerned with how efficient and effective the processes, materials, and workmanship are in producing items. The presumption is that by keeping wastes, rework, and unnecessary operations to a minimum, the cost will be kept in control, which in turn allows the producer to compete with other manufacturers. Customers, on the other hand, tend to concentrate more on the features in a product, including its appearance and convenience to use. Others, such as dealers or consumer groups, however, tend to focus more on the reputation of the product and manufacturer in providing lasting products that perform as promised.

Quality is really multidimensional and to completely define the quality of a product requires specification of a set of attributes that characterizes the item throughout its lifetime. Following Garvin's concept, Thomas (1997) defined product quality as a vector-valued function of the six attributes described in Table 3.1. Performance, the first dimension listed, measures the actual operational performance capabilities of the item providing measures of how well it functions. For example, an automobile or truck engine is rated

TABLE 3.1

Product Quality Dimensions

---
1. **Performance** — Operational performance of the product, such as vehicle fuel consumption in miles per gallon, heater power output in British Thermal Units (BTUs), and the picture resolution in pixels for a digital camera.
2. **Durability** — Ultimate amount of use before the product deteriorates or fails beyond repair.
3. **Reliability** — Probability of product failing within a specified time, having survived to that point.
4. **Conformance** — Degree to which design and operating characteristics comply with preestablished standards.
5. **Aesthetics** — The way in which the product is actually sensed through appearance, feel, sound, touch, and smell.
6. **Perceived Quality** — Overall image of the product among users and potential users.
---

in terms of horsepower, torque, and fuel efficiency. A portable heater might have a heating output of 1000 British Thermal Units (BTUs) per unit of time, and a power washer unit could produce an output of 1500 pounds per square inch (psi) of pressure.

While the capabilities of a product are characterized through its performance measures, the ability to sustain performance and qualities over time is characterized through its durability. Durability is a measure of how long an item will last before it becomes completely nonfunctional. This is measured in time or age of usage, such as miles, cycles, or actual chronological age. Durability is related to and sometimes confused with the reliability of a product. Reliability is a probabilistic statement about the chances of an item surviving over time. It is sometimes expressed as a percentage or even on a subjective category scale, such as low, moderate, and high likelihood. A product can be durable and yet not always be available for service or use due to failures or requirements for periodic maintenance and replacement of parts. Manufacturers, to assess and predict the durability and reliability of products, conduct similar types of aging and destructive testing methods.

For many years product quality had the connotation of being strictly conformance, representing the degree to which standards are met in producing the item. Conformance quality is measured through statistical process control, such as Shewart, statistical process, control charts and other quality control methods for monitoring and tracking process performance over time. Like the other elements, conformance quality is but one of several characteristic elements for quality. The first four dimensions in Table 3.1 can be expressed through objective measures, but the last two, aesthetics and perceived quality, are more difficult to assess, but certainly of no less importance in expressing quality. Aesthetics characterizes how an item appears, through normal sensing of feel, sound, taste, smell, and looks. Perceived quality is the overall image of the product among users and potential users. Brand names tend to have a higher image than an unknown product, based on experience in evaluating different and unrelated product lines. This is an indicator of customer confidence.

3.1.1 Product Life and Quality

The total product life (TPL) can be described in terms of three stages, ranging from the period of inception of its design through its development and effective use as a product (Figure 3.1). The first stage is Product Development, where the item is designed, material selections are made, and the production methods and processes are developed. During this stage some form or level of market analysis is normally conducted to ensure that there will be a demand for the product. The decisions made by designers and technical managers during this stage have greatest impact on durability and reliability. The second stage is Production and Manufacturing, which is the actual production of the product. This includes the operations, scheduling and routing procedures, testing and quality audits, and the distribution to customers and market outlets. To the manufacturer, the dominant dimension at this point is conformance quality, which is assessed and tracked through quality control procedures and inspections. The third stage of the TPL is Product Usage, traditionally considered to be the product life cycle that starts when an item is received by a customer. The TPL is the period that includes the development and production of the product, plus its complete life cycle. The Usage Stage covers the entire span of usage and ownership and includes any resale and extended life through overhaul programs.

3.1.2 The Product System

Let us now consider a product as a system, designed for a particular purpose that will fulfill the needs or requirements of customers and a set of conditions that prescribe the limits of its use. Except for very simple items, a product

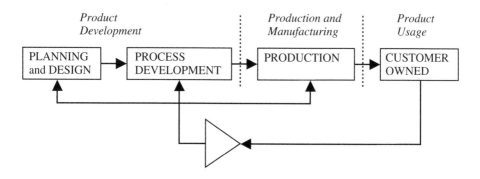

FIGURE 3.1
Stages of total product life.

system consists of a set of subsystems that can be treated as separable and independent elements of the product. Some examples of consumer product systems are (1) an electric toaster, consisting of a cord, heating element, and case that houses the unit; (2) a comfort chair, comprised of a seat, back, two arm rests, foot rest, and a platform; and (3) an automotive vehicle with a body, engine, power train, and interior subsystems. Each subsystem can consist of subsystems and components of hundreds or thousands of parts that are packaged and assembled, often at remote and distant locations, and then transported to a central assembly and distribution site. Many product lines are produced globally with subsystems and components produced in several different countries (e.g., the automobile). As with any system, products can be configured into a number of configurations of components.

3.2 Product Reliability

Reliability is an indication of the amount of failure-free service that can be expected from a product. It is but one of the six quality dimensions in Table 3.1, but it has significant overlapping influence on the other dimensions and TPL. It, therefore, is one of the most vital quality dimensions.

Definition — *Reliability is the probability of an element performing its intended function over its intended life and under specified operating conditions.*

To develop a measure for the reliability of a product or system, it is necessary to establish what constitutes successful performance and the condition of failure, the time period during which successful performance is to be sustained, and the user conditions for the particular item and operational environment. This will all depend on the purpose of the product. Electrical and mechanical devices like switches and light bulbs have simple binary failure and operating states, but most items degrade or deteriorate to an unacceptable level of performance due to such things as noise, vibrations, and controller drift. It is very important that producers know the operating and failure characteristics of their products in order to prescribe the service life and warranty. Producers have some liability for virtually any product through direct or implicit warranties or by law. The conditions of use are also important to communicate to customers to ensure reliability and quality. Statements like "avoid direct sunlight," "use lead-free gasoline only," and "change oil every 3000 miles" are all examples of environmental limitations on products. Specific maintenance programs are sometimes prescribed with conditions to ensure proper care is taken to comply with the operating conditions for a product.

3.2.1 Reliability Measures

There are three fundamental measures of reliability, all derived from the failure characteristics of the item or system that is subject to failure. Let T

Quality and Reliability

be a nonnegative continuous random variable representing the failure time of an item with probability density function f(t) and cumulative distribution F(t), t ≥ 0.

1. *Reliability Function* The most basic measure of reliability is the complement of the failure time distribution, known as the reliability function $R(t) = P(T \geq t)$, or

$$R(t) = 1 - F(t) = \int_t^\infty f(u)du \tag{3.1}$$

   R(t) is shown graphically in Figure 3.2 and represents the probability of an item surviving beyond a time t > 0. We note that $\lim_{t \to 0} R(t) = 1$ and $\lim_{t \to \infty} R(t) = 0$.

2. *Mean Time to Failure (MTTF)* In a perfect setting, one would know the failure characteristics with certainty, but this being too superficial the next best level of risk is to know the failure distribution F(t) and, hence, R(t). Often this is not known exactly, so one resorts to the mean of T

$$E[T] = \int_0^\infty t\, f(t)dt \tag{3.2}$$

and it follows that

$$E[T] = \int_0^\infty R(u)du \tag{3.3}$$

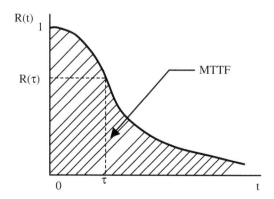

FIGURE 3.2
The reliability function R(t) and MTTF.

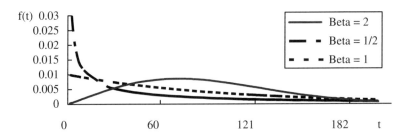

FIGURE 3.3
Weibull pdf ($\theta = 100$; $\beta = 1/2, 1,$ and 2).

This is illustrated in Figure 3.3 by the area under the R(t) vs. t curve. To show this, we consider the right hand side of Equation (3.3) and

$$E[T] = \int_0^\infty [1 - F(t)]dt$$

Integrating by parts, let u = 1-F(t) and dv = dt, therefore du = –f(t)dt and v = t, thus

$$\left([1-F(t)]t\right)\Big|_0^\infty - \int_0^\infty t[-f(t)dt] = \int_0^\infty t f(t)dt$$

3. *Failure Rate and Hazard Function* The reliability function gives the probability of survival from failure beyond some point in time. A related measure is the rate of failures over time, which will generally change throughout the lifetime of a product. Let us start by considering the average failure rate of an item over the time interval (u,u+t) given by

$$G(u, u+t) = \frac{P\{\text{failure in } (u, u+t)|0 \text{ failures by } u\}}{\text{length}(u, u+t)}$$

$$= \frac{P(u < T < u+t | t > u)}{t P(T > u)}$$

It, therefore, follows that

$$G(u, u+t) = \frac{R(u) - R(u+t)}{t \cdot R(u)} \quad (3.4)$$

Quality and Reliability

This gives the average failure rate over a time period, but the more useful measure is the instantaneous rate of failure. This is given by the hazard function, which is derived by

$$\lambda(t) = \lim_{\Delta t \to 0} G(t, t + \Delta t) = \frac{-1}{R(t)} \frac{dR(t)}{dt} \quad (3.5a)$$

from which it follows that

$$\lambda(t) = \frac{f(t)}{R(t)} \quad (3.5b)$$

The hazard function is very important in characterizing failures over time. Another form of R(t) in terms of $\lambda(t)$ can be derived by noting in Equation (3.5a) that we also have

$$\lambda(t) = -\frac{d}{dt} \ln R(t) \quad (3.5c)$$

By multiplying through this equation by dt and integrating, it follows that

$$R(t) = e^{-\int_0^t \lambda(u)\,du} \quad (3.6)$$

Moreover, from Equation (3.5b), $f(t) = \lambda(t)R(t)$ and substituting R(t) from Equation (3.6), we obtain an alternate form of the probablility density function (pdf) given by

$$f(t) = \lambda(t) e^{-\int_0^t \lambda(u)\,du} \quad (3.7)$$

3.2.1.1 Exponential Failure Times

The pdf for the case of T distributed exponential is

$$f(t) = \lambda e^{-\lambda t}, \quad \lambda > 0, \, t \geq 0 \quad (3.8)$$

therefore, from Equation (3.2) we have

$$R(t) = \int_t^\infty \lambda e^{-\lambda u}\,du = e^{-\lambda t} \quad (3.9)$$

and from Equation (3.3)

$$E[T] = \int_0^\infty (1 - e^{-\lambda t}) dt = \frac{1}{\lambda} \quad (3.10)$$

To get the average failure rate, substituting into Equation (3.4)

$$G(u, u+t) = \frac{e^{-\lambda u} - e^{-\lambda(u+t)}}{te^{-\lambda u}} = \frac{1 - e^{-\lambda t}}{t} \quad (3.11)$$

and in Equation (3.5b) we get

$$\lambda(t) = \frac{\lambda e^{-\lambda t}}{e^{-\lambda t}} = \lambda, \; t \geq 0 \quad (3.12)$$

Also note that in Equation (3.11) if the time interval becomes arbitrarily small, $t \to \Delta t$, $e^{-\lambda \Delta t} \approx 1 - \lambda \Delta t$, and $G(u,u+t) \to \lambda$. In other words, if the time interval gets small, the average failure rate approaches the instantaneous failure rate. The exponential distribution is the only failure time distribution with a constant failure rate (CFR). This suggests that failures occur only at random, by chance, and not from some systematic cause.

3.2.1.2 Weibull Failure Times

For T distributed, according to the two-parameter Weibull

$$f(t) = \frac{\beta}{\theta} \left(\frac{t}{\theta}\right)^{\beta-1} e^{-\left(\frac{t}{\theta}\right)^\beta}, \; \beta > 0, \; \theta > 0, \; t \geq 0 \quad (3.13)$$

where β is known as the Weibull slope and θ is called the characteristic value. It follows that the cumulative distribution function (cdf) is

$$F(t) = 1 - e^{-\left(\frac{t}{\theta}\right)^\beta}, \; t \geq 0 \quad (3.14)$$

and, thus, substituting into Equation (3.2) and Equation (3.3) we get

$$R(t) = e^{-\left(\frac{t}{\theta}\right)^\beta}, \; t \geq 0 \quad (3.15)$$

and integrating R(t) over all t ≥ 0

$$E[T] = \theta \Gamma \left(1 + \frac{1}{\beta}\right) \quad (3.16)$$

where

$$\Gamma(x) = \int_0^\infty u^{x-1} e^{-u} du$$

is the incomplete gamma function. Computing the hazard function from Equation (3.5b), we have

$$\lambda(t) = \frac{\frac{\beta}{\theta}\left(\frac{t}{\theta}\right)^{\beta-1} e^{-\left(\frac{t}{\theta}\right)^\beta}}{e^{-\left(\frac{t}{\theta}\right)^\beta}} = \frac{\beta}{\theta}\left(\frac{t}{\theta}\right)^{\beta-1}, \quad t \geq 0 \quad (3.17)$$

Both the exponential and Weibull distributions have wide applications in reliability. The Weibull is a very rich distribution that can take on a broad range of shapes and failure characteristics. These are illustrated in Figure 3.3 and Figure 3.4. Note that $\beta = 1$ in Equation (3.13), the Weibull pdf reduces to the exponential. This is further noted in Figure 3.5 where the failure rate is constant for $\beta = 1$. The other two failure rate curves illustrate that for cases where $0 < \beta < 1$, then $\lambda(t)$ is a decreasing function, corresponding to a decreasing failure rate (DFR) and when $\beta > 1$, it has an increasing failure rate (IFR).

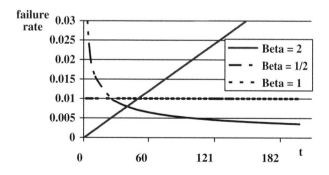

FIGURE 3.4
Failure rates for Weibull ($\theta = 100$; $\beta = 1/2, 1,$ and 2).

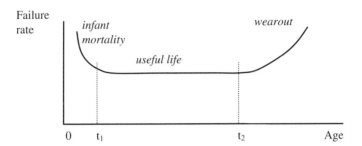

FIGURE 3.5
Product lifetime characteristic curve.

We have provided the developments of the reliability measures for only two distributions, the exponential and Weibull. These measures are summarized in Table 3.2 along with the results for the normal and lognormal as well. These distributions cover many applications and the measures for other distributions can be found in handbooks or derived by applying the results of Equation (3.2) through Equation (3.7).

3.2.2 Failures and Lifetime Characteristics

In this section, we describe the total product lifetime in terms of the various stages that a product transitions throughout its history. The usage stage is what we normally consider as the lifetime and it is characterized by the actual failures that occur over time. Throughout this lifetime, the failure rate will inevitably change as illustrated in the plot of failure rate vs. age shown

TABLE 3.2
Reliability Measures for Some Common Distributions

Distribution	$f(t)$	$R(t)$	$\lambda(t)$	MTTF
Exponential $\lambda > 0, t \geq 0$	$\lambda e^{-\lambda t}$	$e^{-\lambda t}$	λ	$1/\lambda$
Weibull $\beta > 0, \theta > 0,$ $t \geq 0$	$\dfrac{\beta}{\theta}\left(\dfrac{t}{\theta}\right)^{\beta-1} e^{-\left(\frac{t}{\theta}\right)^{\beta}}$	$e^{-\left(\frac{t}{\theta}\right)^{\beta}}$	$\dfrac{\beta}{\theta}\left(\dfrac{t}{\theta}\right)^{\beta-1}$	$\theta\Gamma\left(1+\dfrac{1}{\beta}\right)$
Normal $-\infty < \mu < \infty, \sigma > 0,$ $-\infty < t <$	$\dfrac{1}{\sigma\sqrt{2\pi}} e^{-\frac{1}{2}\left(\frac{t-\mu}{\sigma}\right)^2}$	$1-\Phi\left(\dfrac{t-\mu}{\sigma}\right)$	$\dfrac{f(t)}{R(t)}$	μ
Lognormal $-\infty < \mu < \infty, \sigma > 0,$ $t \geq 0$	$\dfrac{1}{\sigma t\sqrt{2\pi}} e^{-\frac{1}{2}\left(\frac{\ln t-\mu}{\sigma}\right)^2}$	$1-\Phi\left(\dfrac{\ln t-\mu}{\sigma}\right)$	$\dfrac{f(t)}{t\sigma R(t)}$	$e^{\left(\mu+\frac{\sigma^2}{2}\right)}$

TABLE 3.3
Examples of Infant Mortality Failures

Failures	Causes
Seals and welds	Poor workmanship
Solder joints	Material defects
Wire connections	Contamination
Surface scars	Improper material handling
Misalignments	Poor assembly

in Figure 3.5. This is known as the lifetime characteristic curve or, as it is more commonly known, the "bathtub curve." This curve has three distinct periods: infant mortality, useful life, and wearout.

3.2.2.1   Infant Mortality Period

During the very early period of lifetime some, but not all, products experience a high failure rate, but it decreases quite rapidly over a relatively short period. These are sudden failures that generally reflect on the manufacturability of the particular product. Table 3.3 summarizes some examples of failures and related causes for infant mortality. Often times, producers of products known to have inherent infant mortality implement what is known as "burn-in" processes to induce virtual aging to eliminate this period before the items are delivered to customers. This involves cost trade-off decisions, but should include the impact of all of the quality vectors as well, particularly when the reputation of the product is at risk. The infant mortality period is shown by the DFR segment of the bathtub curve from time 0 to t_1.

3.2.2.2   Useful Life Period

Following the burn-in period, product failure rate reaches its lowest and characteristically constant failure rate. The period of useful life, shown in Figure 3.5 from t_1 to t_2, is the period of usage that designers try to achieve when producing lasting products with high durability. Since products have CFR during useful life, the failure times are distributed exponentially and, as mentioned previously, the only failures occur due to randomly occurring chance and not by the way the item was produced and used.

3.2.2.3   Wear-out Period

As a product ages beyond its useful life period (t_2 in Figure 3.5), failures become more frequent due to deterioration of parts and components. Common types of failures during wear out are due to fatigue, creep, and friction wear of mechanical parts; insulation leaks from plastic materials shrinking and cracking; and chemical oxidation on coatings and films. Products in their wear-out period have IFR and maintenance costs become much higher. Decisions for replacement, repair, and overhaul schedules are based on expected wear-out times.

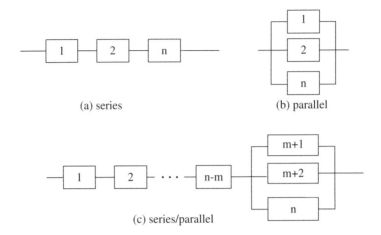

FIGURE 3.6
Basic system configurations.

3.2.3 System Configurations

A product system can consist of several operational subsystems or components that make up the product. Consider a system of n subsystems or components with reliabilities $R_i(t)$, i = 1,...,n. The product system reliability, $R_S(t)$, will depend upon the particular configuration of these subsystems, including the amount of redundancy that is built into the item. The basic configurations are series, parallel, and combined series/parallel arrangements. These are shown in Figure 3.6.

3.2.3.1 Series System

For the series system, failure occurs when any one of the n components fail; therefore, the system reliability is given by

$$R_S(t) = \prod_{i=1}^{n} R_i(t) \tag{3.18}$$

It follows that for a series system, the system reliability will always be worse than the most reliable component. So, as the number of components in series n is increased, $R_S(t)$ will become decreasingly small.

3.2.3.2 Parallel System

When all components are in parallel, as in Figure 3.6b, then system failure will occur only when all n components have failed. The system reliability is then

$$R_S(t) = 1 - \prod_{i=1}^{n}(1 - R_i(t)) \tag{3.19}$$

This is known as simple redundancy and all components are said to be active. Reliability can be increased by adding components in parallel; however, this can be cost-prohibitive when the components are expensive.

3.2.3.3 Combined Series/Parallel System

For most systems, components are added in parallel where it makes sense to add redundancy and can be justified in terms of costs. Figure 3.6c shows a combined system of n components; consisting of a series arrangement of n–m items that is linked to a parallel configuration of the remaining m. The system reliability for this arrangement is

$$R_S(t) = \left[\prod_{i=1}^{n-m} R_i(t)\right]\left[1 - \prod_{i=m+1}^{n}(1 - R_i(t))\right] \quad (3.20)$$

This is a system with built-in redundancy. The choice of components to contain redundancy and the number of components are determined by cost and reliability trade-offs during design.

Exponential Failure Times — Consider the case of the failure times T distributed exponential. For a series system, the reliability function is given by

$$R_S(t) = \prod_{i=1}^{n} e^{-\lambda_i t} = e^{-t\sum_{i=1}^{n}\lambda_i} \quad (3.21)$$

and from Equation (3.5c), the hazard function is

$$\lambda(t) = -\frac{d}{dt}\ln R(t) = \sum_{i=1}^{n}\lambda_i \quad (3.22)$$

Integrating Equation (3.21), we obtain the MTTF

$$E[T] = \int_0^\infty e^{-\sum_{i=1}^n \lambda_i t}\,dt = \left.\frac{e^{-t\sum_{i=1}^n \lambda_i}}{-\sum_{i=1}^n \lambda_i}\right|_0^\infty$$

or

$$E[T] = \frac{1}{\sum_{i=1}^{n}\lambda_i} \quad (3.23)$$

For a parallel arrangement of subsystems with exponential distributed failure times, the reliability function from Equation (3.19) is

$$R_S(t) = 1 - \prod_{i=1}^{n}[1 - e^{-\lambda_i t}] \qquad (3.24)$$

Note here that though the subsystem failure times are distributed exponentially, when the subsystems are in parallel the resulting system failure time distribution is not exponential. Suppose we have n = 3 subsystems. The system reliability function is then

$$R_S(t) = 1 - (1 - e^{-\lambda_1 t})(1 - e^{-\lambda_2 t})(1 - e^{-\lambda_3 t})$$

which simplifies to

$$\begin{aligned}R_S(t) = & e^{-\lambda_1 t} + e^{-\lambda_2 t} + e^{-\lambda_3 t} - e^{-(\lambda_1+\lambda_2)t} - e^{-(\lambda_2+\lambda_3)t} \\ & - e^{-(\lambda_1+\lambda_3)t} + e^{-(\lambda_1+\lambda_2+\lambda_3)t}\end{aligned} \qquad (3.25)$$

and the MTTF is

$$E[T] = \int_0^{\infty} (e^{-\lambda_1 t} + e^{-\lambda_2 t} + e^{-\lambda_3 t} - e^{-(\lambda_1+\lambda_2)t} - e^{-(\lambda_2+\lambda_3)t} - e^{-(\lambda_1+\lambda_3)t} + e^{-(\lambda_1+\lambda_2+\lambda_3)t}) dt$$

which leads to

$$E[T] = \frac{1}{\lambda_1} + \frac{1}{\lambda_2} + \frac{1}{\lambda_3} - \frac{1}{\lambda_1+\lambda_2} - \frac{1}{\lambda_2+\lambda_3} - \frac{1}{\lambda_1+\lambda_3} + \frac{1}{\lambda_1+\lambda_2+\lambda_3} \qquad (3.26)$$

Examples

1. Three components are connected in parallel with failure times that are distributed exponential with MTTF of 20, 50, and 100 hours, respectively. Determine the system reliability for 10 hours of operation and MTTF.

SOLUTION:
The exponential parameters follow from Equation (3.23),

$$\lambda_i = 1/\text{MTTF}_i$$

for i = 1, 2, 3 and, hence, $\lambda_1 = 0.05$, $\lambda_2 = 0.02$, and $\lambda_3 = 0.01$. From Equation (3.25), the reliability function is

$$R_S(t) = e^{-0.05t} + e^{-0.02t} + e^{-0.01t} - e^{-(0.05+0.02)t}$$

$$- e^{-(0.02+0.01)t} - e^{-(0.05+0.01)t} + e^{-(0.05+0.02+0.01)t}$$

$$= e^{-0.05t} + e^{-0.02t} + e^{-0.01t} - e^{-0.07t} - e^{-0.03t} - e^{-0.06t} + e^{-0.08t}$$

and for t = 10 hours,

$$R_S(10) = 0.9931.$$

The MTTF for the system from Equation (3.26) is

$$E[T] = \frac{1}{.05} + \frac{1}{.02} + \frac{1}{.01} - \frac{1}{.07} - \frac{1}{.03} - \frac{1}{.06} + \frac{1}{.08} = 118.2 \text{ hrs}$$

2. Consider radar comprised of a transmitter, receiver, antenna, and controls package subsystems. This radar cannot function without each of these subsystems. The system is therefore a series configuration as shown in Figure 3.7. The system is assumed to be functioning during its useful life with the following failure rates:

Subsystem	Description	Failure Rate (fr/hr)	Cost ($ × 1000)
1	Transmitter	2.3 × 10⁻⁶	75
2	Receiver	8.5 × 10⁻⁵	45
3	Antenna	1.2 × 10⁻⁶	70
4	Controls Package	4.7 × 10⁻⁵	22

a. Determine the reliability and hazard function of the radar for a 2000-hour operational period.

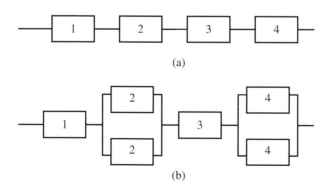

FIGURE 3.7
Example of radar system with (a) series, (b) parallel, and combined series/parallel configurations.

b. Compute the reliability for a 2000-hour period if the system is modified to have a redundant receiver and controls package subsystems.

SOLUTION:

a. Since the radar and, therefore, the subsystems are all functioning during useful life, the failure times are distributed exponential; therefore, from Equation (3.9)

$$R_1^{(a)}(2,000) = \exp[-(2.3 \times 10^{-6})(2000)] = 0.9954$$

and similarly $R_2^{(a)}(2000) = 0.8437$, $R_3^{(a)}(2000) = 0.9976$, and $R_4^{(a)}(2000) = 0.9103$ and it follows from Equation (3.18) that $R_S^{(a)}(2000) = 0.7626$. From Equation (3.21), the hazard function is $\lambda(t) = (2.3 + 85 + 1.2 + 47) \times 10^{-6} = 1.355 \times 10^{-6}$ failures per hour.

The cost for this system is

$$K^{(a)} = K_1 + K_2 + K_3 + K_4 = 75 + 45 + 70 + 22 = \$212,000$$

b. To improve the reliability by building in complete redundancy for subsystems 2 and 4, the new system is as shown in Figure 3.7b. The reliability for the parallel segments are

$$R_2^{(b)}(2000) = 1 - \left[1 - R_2(2000)\right]^2 = 0.9756$$

and

$$R_4^{(b)}(2000) = 1 - \left[1 - R_4(2000)\right]^2 = 0.9919$$

The system reliability for this arrangement then is

$$R_S^{(b)}(2000) = (0.9954)(0.9756)(0.9976)(0.9919) = 0.9610$$

and the cost increases to

$$K^{(b)} = K_1 + 2K_2 + K_3 + 2K_4 = 75 + 2(45) + 70 + 2(22) = \$279,000$$

During useful life where the failure times are distributed exponential the reliability measures are quite simple. When T has some other distribution, R(t) and $\lambda(t)$ can become quite cumbersome but can be determined by applying Equation (3.18) through Equation (3.20).

3.2.4 Reliability Improvement and Redundancy

Most product manufacturers are engaged in some type of continuous quality improvement program. Reliability is, of course, a primary focus area for making improvements. There are three avenues for improving the reliability of a product system: (1) improvement of component level reliability, (2) implementation of new technology, and (3) increased redundancy. The first avenue is to identify opportunities for improvement of the design and methods throughout the system. There is normally a learning curve that follows the implementation of any new system or when major changes in the processes and materials used in its production and distribution are implemented. The second option to consider is whether it is feasible or cost effective to replace components and subsystems by more advanced systems that will reduce or eliminate failure modes and provide greater feedback for control through improved sensing and diagnostics capabilities. Electronic ignition, fuel injection, and in-process sensing and warning systems are examples of technological improvements. The third option for improving the reliability is to add additional components in parallel where it makes sense. These decisions involve cost trade-offs with R(t) as computed using Equation (3.19) and Equation (3.20).

Quite often it is neither practical nor necessary to build in redundancy by completely duplicating or adding multiples of the components in parallel where every component is in an active status when the system is online. Two common forms of improvement through redundancy that overcome the need for as many active components are *standby redundancy* and *load sharing*. Standby redundancy consists of connecting additional units that are held in a standby status and then activated through a switching mechanism upon the failure of an active unit that is online. Auxiliary power supply units used to back up power supplied to hospitals and other critical systems requiring high reliability for life support are examples of standby redundancy.

In load-sharing applications, all units are active, but whenever a unit fails, the surviving elements receive an increase in their load. The lug bolts used to secure the load on an automobile or truck wheel form a load sharing system. If one bolt shears off, the remaining ones can sustain the load by each assuming an increase in their share.

Reliability models and methods for incorporating standby redundancy and load sharing can be found in basic reliability texts, such as Ebeling (2005).

3.3 Load and Capacity Models

We define a loading as an event whereby an element having a capacity C is subjected to a load L, as shown in Figure 3.8. A failure occurs whenever, for a particular loading, the interference, I = C−L<0. We will develop results for computing the failure probability,

$$p = P(I < 0), \quad 0 < p < 1 \qquad (3.27)$$

for a loading with the understanding that the associated reliability is simply r = 1−p. Either L or C or both can be random variables. For the case where

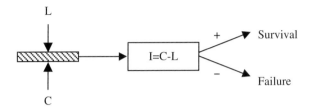

FIGURE 3.8
Schematic of a failure loading condition.

L is distributed $F_L(y)$ with constant capacity c, the probability of a loading failure is

$$p = P(I<0) = P(L>c) = \int_c^\infty f_L(y)\,dy$$

or

$$p = 1 - F_L(c) \qquad (3.28)$$

For C distributed G_C with the load l constant

$$P = P(I<0) = P(C<l) = G_C(l) \qquad (3.29)$$

For the case of both L and C random variables with failure time distributions $F_L(x)$ and $G_C(y)$, respectively, by conditioning on L the failure probability is given by

$$p = \int_0^\infty P(C<y|y<L<y+dy) f_L(y)\,dy$$

or

$$p = \int_0^\infty G_C(y) f_L(y)\,dy \qquad (3.30)$$

Alternatively, by conditioning on C, it follows that

$$p = \int_0^\infty R(y) g_C(y)\,dy \qquad (3.31)$$

Example 3.1 (from Rao, 1992)

The circumferential stress that builds up in a thin-walled cylindrical pressure vessel for a particular application is

$$S_\theta = \frac{\Pi d}{2\tau} \tag{3.32}$$

where d is the inside diameter of the cylinder, τ is the wall thickness, and Π is the internal pressure that is exponentially distributed with a mean of 400 psi. The yield strength of the vessel, S_y, is exponentially distributed with a mean of 30,000 psi. Suppose the parameter values are d = 72 in. and τ = 0.5 in.. Assume that S_y is also distributed exponential with mean 400 psi. We want to determine the reliability of this vessel.

SOLUTION:
Here we have the capacity C is the yield strength S_y, which is distributed exponential with parameter $\lambda_C = 1/30,000$, and the load is the stress given by

$$L = S_\theta = \Pi \frac{72}{2(0.5)} = 72\Pi$$

and, since Π is distributed exponential, L is also distributed exponential with a mean of

$$\lambda_L = 1/[(400)(72)]$$

In general, for L and C distributed exponential from Equation (3.29), the failure probability is

$$p = \int_0^\infty [1 - e^{-\lambda_L y}] \lambda_C e^{-\lambda_C y} dy = 1 - \int_0^\infty \lambda_C e^{-(\lambda_L + \lambda_C) y} dy \tag{3.33}$$

$$= \frac{\lambda_C}{\lambda_L + \lambda_C}$$

therefore, the reliability is

$$r = \frac{\lambda_L}{\lambda_L + \lambda_C} \tag{3.34}$$

Substituting the values for λ_L and λ_C, we have

$$r = \frac{\frac{1}{(72)(400)}}{\frac{1}{(72)(400)} + \frac{1}{30,000}} = 0.5102$$

Loadings can arise through a variety of sources, depending on the products and conditions of the operating environment. Equation (3.28) through Equation (3.31) give the point failure probabilities for a single loading, such as a punch press engaging a metal work piece, a surge through a circuit breaker, or the activation of a switch. Loadings can be periodic or at random intervals, both cases resulting in random failure times.

3.3.1 Periodic Loadings

Consider the case where loadings occur periodically in fixed cycles $t = 1,2,\ldots$, at which time failure occurs with constant probability p, as determined from Equation (3.28) through Equation (3.31). It follows that the discrete probability function of the time to failure T is geometric with

$$p_T = P(T = t) = pq^{t-1}, \quad q = 1 - p, t = 1, 2, \ldots \quad (3.35)$$

with the cumulative distribution

$$F(t) = P(T \le t) = \sum_{x=1}^{t} pq^{x-1} = p\frac{1-q^t}{1-q}$$

or

$$F(t) = 1 - (1-p)^t, \quad t = 1, 2, \ldots \quad (3.36)$$

The reliability function is then

$$R(t) = (1 - p)^t$$

and, for the case of L and C random, substituting Equation (3.30) for p leads to

$$R(t) = \left(1 - \int_0^\infty G_C(y) f_L(y) dy\right)^t, \quad t = 1, 2, \ldots \quad (3.37)$$

Alternatively, applying the version of Equation (3.31) for the loading failure probability, p results in

$$R(t) = \left(1 - \int_0^\infty R(y) g_C(y) dy\right)^t, \quad t = 1, 2, \ldots \quad (3.38)$$

Example 3.2

Suppose the pressure vessel in Example 3.1 receives the loading on a daily basis, say every day at 3 p.m. We already computed the loading reliability

Quality and Reliability 55

from Equation (3.33) and the failure probability at each loading is p = 1 − r = 1 − 0.5102 = 0.4898.

The reliability function is then

$$R(t) = (0.5102)^t, \quad t = 1, 2, \ldots$$

To get the MTTF, we use the discrete analog to Equation (3.3), since t is integer valued,

$$\text{MTTF} = \sum_{t=1}^{\infty} R(t) = \sum_{t=1}^{\infty} (0.5102)^t = \frac{0.5102}{1 - 0.5102} = 1.042 \text{ days}$$

3.3.2 Random Loadings

Now consider the case of the loadings occurring at completely random times T. By completely random, we mean the instantaneous probability of a loading event occurring is a constant $\lambda > 0$ and for T being a continuous random variable, the failure time distribution is exponential. It then follows that the number of such loadings N(t) in an interval (0,t] is distributed Poisson with rate λ. (see Section 2.2.3). Extending this further, the conditional probability of a failure given N(t) = n is distributed geometric with parameter p. The reliability function is then found by

$$R(t) = \sum_{n=1}^{\infty} P(\text{failure} \mid n \text{ loadings}) P[N(t) = n]$$

or

$$R(t) = \sum_{n=1}^{\infty} pq^{n-1} \frac{(\lambda t)^n}{n!} e^{-\lambda t} = \frac{pe^{-\lambda t}}{q} \sum_{n=1}^{\infty} \frac{(q\lambda t)^n}{n!}$$

$$= \frac{pe^{-\lambda t}}{q} \left[(q\lambda t) + \frac{(q\lambda t)^2}{2!} + \frac{(q\lambda t)^3}{3!} + \cdots \right] = p\lambda t e^{-\lambda t} \cdot e^{q\lambda t}$$

therefore,

$$R(t) = p\lambda t e^{-p\lambda t}, \quad t \geq 0 \tag{3.39}$$

Example 3.2 (continued)

Suppose the loadings on the pressure vessel occur at times that are distributed exponential with a mean of 30 operating hours. With a failure probability at each loading of p = 0.4898, the reliability function is

$$R(t) = (0.4898)(1/30) t e^{-(0.3898)(1/30)t}$$

or

$$R(t) = 0.1633te^{-0.1633t}, \quad t \geq 0$$

and the mean time to failure from Equation (3.3) is

$$\text{MTTF} = \int_0^\infty 0.1633te^{-0.1633t}dt = \frac{1}{0.1633} = 6.12 \text{ hours}$$

Reliability interference is the standard approach for modeling the physical characteristics of a failure process. Further details and derivations of these models are given in Kapur and Lamberson (1977), including cases where the load and capacity distributions are different.

3.4 Exercises

1. Another important measure used in reliability design is the *design life*, defined as the time t_R that will provide a prescribed reliability level R^*. Derive the design reliability for each of the following:
   a. Normal
   b. Two-parameter Weibull

2. Given the failure time distribution for a subsystem is Weibull, with the probability density function

$$f(t) = \frac{\beta}{\theta}\left(\frac{t}{\theta}\right)^{\beta-1} e^{-(t/\theta)^\beta}, \quad \beta > 0, \ \theta > 0, \ t \geq 0$$

   a. Derive the MTTF and compute Var[T].
   b. Determine the 0.95 reliability for the case of $\beta = 2$ and $\theta = 100$ hours.
   c. Suppose a product consists of three of these subsystems in series, find the system reliability for an operating time of 200 hours.

3. Let T be distributed according to a Weibull with parameters $\beta > 0$ and $\theta > 0$.
   a. Verify that

$$\text{MTTF} = \int_0^\infty R(u)du$$

b. Show that the median time to failure is less than the characteristic lifetime $\theta > 0$.

4. The conditional reliability is the reliability at time $t > 0$, given a component has functioned for t_0 units of time
   a. Show that
   $$R(t|t_0) = e^{-\int_{t_0}^{t+t_0} \lambda(\tau)d\tau}$$

   b. Consider a device with the failure time T distributed
   $$F(t) = 1 - e^{-(t/\theta)^\beta}, \quad \beta > 0, \ \theta > 0, \ t \geq 0$$

   Compute the probability of a failure during the second year of a 2-year warranty, given that no failure occurred in the first year.

5. The failure time for a component is distributed gamma with probability density function
   $$f(t) = \frac{\lambda(\lambda t)^{r-1}}{\Gamma(r)} e^{-\lambda t}, \quad t \geq 0, \ \lambda > 0, \ r = 1, 2, \ldots$$

   Derive the reliability and hazard functions.

6. A component that is crucial to the operation of a given system has failure time T_1 that is distributed exponential with MTTF of $1/\lambda$. Upon failure, this component is replaced by an auxiliary unit that is automatically energized online with failure time T_2 that has the probability density function
   $$f(t) = \frac{\lambda^3 t^2}{2} e^{-\lambda t}, \quad t \geq 0$$

   a. Find the probability density function for the component total time to failure $T = T_1 + T_2$.
   b. Give the MTTF and establish whether the distribution for T is IFR or DFR.
   c. Given $\lambda = 0.05$ failures/hour, compute the reliability for an operating period of 15 hours, starting at an age of 135 hours.

7. An appliance manufacturer produces an electric iron that is sold under a warranty that provides replacement for any failure that occurs during the first year. The failure time is distributed Weibull

with a characteristic value of 3 years and shape parameter $\beta = 2$. Compute the reliability for the warranty period.

8. The reliability for a particular cutting assembly can be approximated by

$$R(t) = \begin{cases} (1-t/t_0)^2, & 0 \leq t \leq t_0 \\ 0, & t \geq t_0 \end{cases}$$

   a. Determine the instantaneous failure rate.
   b. Show that the failure rate increases or decreases.
   c. Determine the MTTF.

9. The failure time in operating hours of a power unit has the hazard function

$$\lambda(t) = 0.003\sqrt{t/500}, \, t \geq 0$$

   a. Compute the reliability for 50 hours of operation.
   b. Determine the design life for a reliability goal of 0.90.
   c. Find the reliability of 10 hours of use given the unit has been in operation for 200 hours.

10. Consider a series arrangement of three components A, B, and C, each having reliability R. Show that for the options to increase the system reliability, R_S, through high level (H) and low level (L) redundancy, $R_L > R_H$.

11. Show that for a device that is subject to a single load L that is distributed F(x) and capacity C distributed G(y), the failure probability is given by Equation (3.31).

References

Ebeling, C.E., 2005, *An Introduction to Reliability and Maintainability Engineering*, Waveland Press, Long Grove, IL.

Garvin, D.A., 1987, Competing on the eight dimensions of quality, *Harv. Busn. Rev.*, 65, 6: 107–119.

Kapur, K.C. and L.R. Lamberson, 1977, *Reliability in Engineering Design*, John Wiley & Sons, New York.

Rao, S.S., 1992, *Reliability-Based Design*, McGraw-Hill, New York.

Thomas, M.U., 1997, A methodology for product assessment, *Productivity and Quality Management Frontiers*, Vol. 6; C.G. Thor, J.A. Edossomwon, R. Poupart, and D.J. Sumanth, Eds., Engineering and Management Press, Norcross, GA, pp. 649-659.

4

Economic Models for Product Warranties

In Chapter 3, we defined product quality as a multidimensional function of the performance, durability, reliability, conformance, aesthetics, and perceived quality of a product. These attributes establish the value of an item to a customer in terms of its design, how it is produced, and how it appears as an overall product system. Whenever a product is considered to be of low quality, it could be due to any number of causes. The relative importance of the quality attributes will vary among different people, particularly between manufacturers and consumers. Therefore, it is difficult to come up with simple measures for overall quality. However, it is clear that every element of the product quality vector will impact the warranties that are provided to consumers.

The origin of warranties is not known, but there seems to be historical evidence of their existence as early as 2000 B.C. They, no doubt, have been around as long as people have been engaged in the exchange of goods and services. A complete historical perspective and survey of warranty liability is given in Blischke and Murthy (1996). In this chapter we will describe the fundamental warranty policies that are applied to consumer products and present the cost models for dealing with warranty economic decisions.

4.1 Definitions and Types of Warranties

Definition — *A warranty is a formal commitment to customers by a producer of products or services to assume certain responsibilities for the quality of the units following sale or delivery.*

During the warranty period, the producer assumes some portion of the expenses that result from the repair or replacement of products that have become defective. This provides customers with relief from the financial risk of low quality during the warranty period. When the quality of a product is high, then the manufacturer's cost for the warranty will be relatively low. On the other hand, when products of low quality are released to the market, eventually the manufacturer will realize increased warranty expenses.

Consequently, the warranty program must be an integral part of quality management and continuous improvement.

To specify a warranty policy requires a set of conditions for the compensation for failures and malfunctions, and a period of coverage. The factors involved in establishing a particular policy include the product failure characteristics, expected impact on customer preferences and reactions, and the marginal costs of warranty claims. While there are many types of warranties and several ways of classifying them, they are all based on free replacement or pro rata rebate schemes.

Let T represent the time of failure with distribution F(t), and X_i the associated cost to the manufacturer for a failure that occurs during the warranty period (0,w).

4.1.1 Free-Replacement Warranty (FRW)

Under a free replacement policy, the manufacturer absorbs all costs associated with the repair or replacement of products that fail during the warranty period. The FRW policy is common for repairable items like televisions, washing machines, and lawn mowers that can be returned to a good-as-new operating condition through repairs. It also applies to nonrepairable items, such as flash bulbs, disposable medical products, and inexpensive electronic components.

Let T be a random variable representing the time to failure of a product that has a repair or replacement cost to the manufacturer of c_1 dollars per unit. Under the FRW policy, the unit cost to the manufacturer at a time $t \geq 0$ t for a warranty of length w > 0, shown in Figure 4.1a, is

$$X(t) = \begin{cases} c_1, & 0 < t \leq w \\ 0, & t > w \end{cases} \quad (4.1)$$

Therefore, the expected cost of a single failure is

$$E[X(t)] = \int_0^w c_1 f(t) dt$$

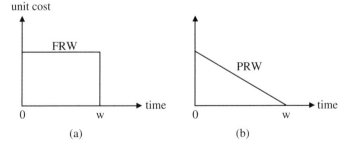

FIGURE 4.1
Unit warranty cost for FRW and PRW policies.

or

$$E[X(t)] = c_1 F(w) \tag{4.2}$$

where f(t) and F(t) are the probability density and cumulative density functions for T.

The FRW policy is commonly applied to products that are not repairable or repairable items, such as televisions, washing machines, and lawn mowers that can be brought back to good-as-new condition through standard repairs.

Example 4.1

An appliance manufacturer produces a particular model of toaster at a cost of $15 per unit and sells it with a 1-year warranty on all parts and labor. The failure rate for the toaster is 0.0025 failures per operating hour and it is assumed to function during its useful life.

Since the item is functioning during its useful life, the failure time distribution is exponential and the parameter is λ = 0.0025 failures per hour. Assuming that a consumer uses the toaster 100 hours per year, then λ = 0.25 failures per year. With a cost to the manufacturer of c = $15, the average unit cost during a 1-year warranty is

$$E[X(t)] = 15 F(1) = 15 \left[1 - e^{-(0.25)(1)} \right] = \$3.32$$

4.1.2 Pro Rata Replacement Warranty (PRW)

A pro rata policy provides customers with an amount of rebate that is a function of the age of the item when it fails within the warranty period. This policy is applied to nonrepairable items. Products like tires are generally sold with a PRW policy.

Under a PRW policy with a linear pro rata function, an item that costs the manufacturer c_0 dollars per unit with a warranty of length w > 0 and has a unit cost at time t, as shown in Figure 4.1b, is given by

$$X(t) = \begin{cases} c_0 \left(\dfrac{w-t}{w} \right), & 0 < t \leq w \\ 0, & t > w \end{cases} \tag{4.3}$$

The customer must be willing to pay the complement portion of the cost, i.e., $c_0 t / w$, to execute the warranty under a PRW policy. The expected unit cost is

$$E[X(t)] = \int_0^w c_0 \left(\frac{w-t}{w} \right) f(t) dt$$

or

$$E[X(t)] = c_0 F(w) - \frac{c_0}{w} \int_0^w t f(t) dt \qquad (4.4)$$

Example 4.2
A turbine engine with a manufacturer's cost of $1000 is sold under a 15,000-hour PRW policy. The failure rate for the engine is 2×10^{-5} hours of operation. Assume the engine functions during its useful life. Find the expected unit warranty cost.

SOLUTION
Substituting into Equation (4.4) for the exponential distribution

$$E[X(w)] = c_0[1 - e^{-\lambda w}] - \frac{c_0}{w} \int_0^w t \lambda e^{-\lambda t} dt \qquad (4.5)$$

$$= c_0(1 - e^{-\lambda w}) - \frac{c_0}{\lambda w}[1 - (1 + \lambda w)e^{-\lambda w}]$$

For the case where $c_0 = \$1,000$, $w = 15,000$ hours, and $\lambda = 2 \times 10^{-5}$ hours,

$$E[X(15,000)] = 1000[1 - e^{-(2x10^{-5})(15000)}] - \frac{1000}{(2x10^{-5})(15000)}$$

$$\times \{1 - [1 + (2x10^{-5})(15,000)]e^{-(2x10^{-5})(15000)}\}$$

$$= (1000)(0.25918) - (3333.33)\{1 - (1.3)(0.74082)\}$$

$$= 259.18 - 123.12 = \$136.06$$

4.1.3 Warranty Costs and Discounting

When economic conditions are such that the time value of money cannot be ignored, it becomes necessary to discount the costs that accrue when analyzing and evaluating warranty decisions. This will always be the case for products that have high repair or replacement costs and lengthy warranties. Denoting the discounting factor by δ, $0 < \delta < 1$, the discounted expected unit costs are computed as follows:

a. FRW Policy — The unit cost at time t is

$$X(t) = \begin{cases} c_1 e^{-\delta t}, & 0 < t \leq w \\ 0, & t > w \end{cases} \qquad (4.6)$$

therefore, the expected cost is given by

$$E[X(t)] = c_1 \int_0^w c_1 e^{-\delta t} f(t) dt \qquad (4.7)$$

For the case of exponential distributed failures,

$$E[X(t)] = c_1 \int_0^w e^{-\delta t} \lambda e^{-\lambda t} dt = \lambda c_1 \int_0^w e^{-(\lambda+\delta)t} dt$$

or

$$E[X(t)] = \frac{\lambda c_1}{\lambda + \delta} \left[1 - e^{-(\lambda+\delta)w} \right]. \qquad (4.8)$$

Suppose in Example 4.1, the appliance manufacturer applies discounting with $\delta = 0.05$.

$$E[X(t)] = \frac{(15)(.25)}{(.25+.05)} [1 - e^{-(.25+.05)1}] = \$3.24$$

b. PRW Policy — The unit cost at time t is

$$X(t) = \begin{cases} c_1 e^{-\delta t} \left(\dfrac{w-t}{w} \right), & 0 < t \leq w \\ 0, & t > w \end{cases} \qquad (4.9)$$

and

$$E[X(t)] = c_1 \int_0^w e^{-\delta t} \left(1 - \frac{t}{w} \right) f(t) dt$$

$$= c_1 \int_0^w e^{-\delta t} f(t) dt - \frac{c_1}{w} \int_0^w e^{-\delta t} t f(t) dt$$

It also follows by integrating the right-hand side by parts that

$$E[X(t)] = \frac{c_1}{w} \int_0^w \int_0^t e^{-\delta \tau} f(\tau) d\tau \qquad (4.10)$$

With this PRW policy and exponential distributed failure times, we have

$$E[X(t)] = \frac{c_1}{w} \int_0^w \int_0^t \lambda e^{-(\delta+\delta)\tau} d\tau$$

$$= \frac{c_1}{w} \int_0^w \frac{\lambda(1 - e^{-(\lambda+\delta)t})}{\lambda + \delta} dt$$

which simplifies to

$$E[X(t)] = \frac{\lambda c_1}{\lambda + \delta} - \frac{\lambda c_1}{w(\lambda + \delta)^2}[1 - e^{-(\lambda+\delta)w}] \qquad (4.11)$$

Applying a discount factor of 5% to the turbine engine in Example 4.2, we first need to convert the discount factor, which is implicitly given on a per year basis to hourly units. Assume the engine is used 10 hours per day for 250 work days per year, therefore

$$\delta = 0.05 \times 10 \text{ hours/day} \times 250 \text{ days/year} = 0.00002$$

and

$$\lambda + \delta = 0.00002 + 0.00002 = 0.00004$$

Thus, we have

$$E[X(t)] = \frac{(1000)(.00002)}{.00004} - \frac{(1000)(.00002)}{(15,000)(.00004)^2}\left[1 - e^{-(.00004)(15,000)}\right]$$

$$= 500 - (833.33)(.4) = \$166.67$$

4.1.4 Combined FRW/PRW

Many products are sold under a combined policy whereby items that fail during an initial period $(0, w_1)$ are covered in full by an FRW policy, followed by a PRW policy during (w_1, w_2). Let c_0 be the initial cost. The unit cost, as shown in Figure 4.2, is given by

$$X(t) = \begin{cases} c_0, & 0 < t \le w_1 \\ c_0\left(\dfrac{w_2 - t}{w_2 - w_1}\right), & w_1 < t \le w_2 \\ 0, & t > w_2 \end{cases} \qquad (4.12)$$

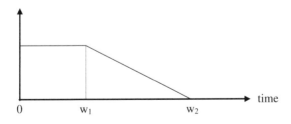

FIGURE 4.2
Unit warranty cost for a combined FRW/PRW policy.

Economic Models for Product Warranties

The expected cost for a single claim, i.e., single failure, is therefore,

$$E[X(t)] = c_0 \int_0^{w_1} f(t)\,dt + \frac{c_0}{w_2 - w_1} \int_{w_1}^{w_2} (w_2 - t) f(t)\,dt$$

$$= c_0 F(w_1) + \left(\frac{c_0 w_2}{w_2 - w_1}\right)[F(w_2) - F(w_1)] - \frac{c_0}{w_2 - w_1} \int_{w_1}^{w_2} t f(t)\,dt$$

Integrating the last term by parts and simplifying,

$$E[X(t)] = \left(\frac{c_0}{w_2 - w_1}\right) \int_{w_1}^{w_2} F(t)\,dt \tag{4.13}$$

Example 4.3

A tire manufacturer produces a particular brand of tire at a cost of $50 per tire that is sold under a 24-month warranty with a 90-day FRW policy followed by a PRW policy for the remaining period. Failure times are distributed exponential with a MTTF of 30 months. Assume customers are willing to pay their prorated portion of replacement costs when it is required.

Here we have the parameter values: $c_0 = \$50$, $w_1 = 3$ months, $w_2 = 24$ months, $\lambda = 1/30$.

Substituting into Equation (4.7), the expected unit cost is

$$E[X(24)] = \left(\frac{50}{24 - 3}\right) \int_3^{24} \left(1 - e^{-t/30}\right) dt$$

$$= 2.38\left[21 + 30\left(e^{-24/30} - e^{-3/30}\right)\right] = \$17.46$$

Suppose the manufacturer wants to develop a marketing option for customers to select a complete FRW policy as an option to the combined FRW/PRW policy. This optional FRW policy, of course, would have to be over a shorter period, $(0, w_F)$ with $w_F < w_2$, unless the manufacturer is willing to incur a greater cost. Thomas (1981) showed that an *economically equivalent* FRW policy can be derived by equating the unit failure costs given in Equation (4.2) and Equation (4.13),

$$cF(w_F) = \left(\frac{c}{w_2 - w_1}\right) \int_{w_1}^{w_2} F(t)\,dt$$

We let $c_0 = c_1 = c$ and it follows that

$$w_F = F^{-1}\left(\int_{w_1}^{w_2} \frac{F(t)}{w_2 - w_1}\,dt\right) \tag{4.14}$$

For the case of exponentially distributed failure times, it is easily shown that

$$w_F = \frac{1}{\lambda} \ln\left[\frac{\lambda(w_2 - w_1)}{e^{-\lambda w_1} - e^{-\lambda w_2}}\right] \quad (4.15)$$

So, with $\lambda = 1/30$, $w_1 = 3$ and $w_2 = 24$ months

$$w_F = 30 \ln\left[\frac{(1/30)(24-3)}{e^{-3/30} - e^{-24/30}}\right] = 12.9 \text{ months}$$

Therefore, the manufacturer can offer an FRW policy over (0,13) months that will result in an average cost that is no greater than that of the combined FRW/PRW policy described in Example 4.3.

The duration of coverage of a warranty can be fixed or it can be renewable. For a fixed period of, say, 1 year, the product is covered only for that period and expires at the end of the year irrespective of the number of warranty claims filed during the year. Renewable period warranties start with a fixed interval, but each failure within the warranty interval results in a new warranty (i.e., the warranty is renewed with a new fixed interval). Therefore, the actual length of coverage for a renewing warranty is random. Renewable period warranties are normally used for products having very high quality and are usually inexpensive. FRW policies can be either fixed or renewable period, but PRW policies are generally fixed period.

4.2 Warranty Cost Models

The total costs consist of the initial cost of the item plus the accumulation of the costs for failures throughout the warranty period. However, for non-renewing warranties, the period of coverage is fixed, but for renewing warranties the actual period of coverage will depend on the failure times. This can be seen by examining the sample realization of the counting process for failures under a renewing type of warranty in Figure 4.3. Under a regular

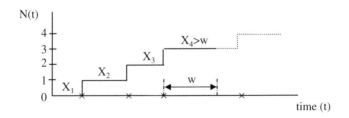

FIGURE 4.3
Sample realization of $\{N(t): t \geq 0\}$ for a renewing warranty.

Economic Models for Product Warranties

nonrenewing warranty, the coverage expires after a fixed period of time w, and if a failure occurs at a time t < w, then the repaired or replaced item is covered only for the residual period (t, w). For a renewing warranty, however, the actual coverage continues until the first incidence at which the time between failures exceeds length w. So, in theory, the effective warranty period of coverage could be indefinite.

4.2.1 Nonrenewing Warranty

Consider a product that is produced at an initial cost to the manufacturer of c_0 and put into service at, say, time 0 under a warranty policy of length w > 0. We define random variables $X_i(t)$ as the unit cost due to failure i at time t where i = 1,2,..., and N(t) the number of failures in (0,t). The total cost is then

$$Y(w) = c_0 + \sum_{i=1}^{N(w)} X_i(t) \tag{4.16}$$

Since the times between failures are independent and identically distributed F(t) and N(w) is a stopping time, then it follows from Wald's Equation (see Ross, 1983) that the expected total cost is given by

$$C(w) = E[Y(w)] = c_0 + E[X_i(w)]M(w) \tag{4.17}$$

where

$$M(t) = \sum_{n=1}^{\infty} F^{(n)}(t)$$

is the Renewal Function and $F^{(n)}(t)$ is the n-fold convolution of F with itself, defined in Chapter 2, Section 2.3. For many, if not most, applications, $E[X_i(w)]$ is treated as a constant average unit cost, thus allowing the simpler expected cost function

$$C(w) = c_0 + c_1 M(w) \tag{4.18}$$

Except for a few special cases, such as the binomial and Poisson counting processes, M(t) does not have a simple closed form. One such case is when products function during their useful life and, hence, the failure times are exponentially distributed. M(t) is then a simple linear function of time. It turns out that for most consumer products the number of failures that occurs during the warranty period rarely exceeds one. Otherwise the product has significant quality problems and it probably was prematurely placed on the market. We will call this the *single-failure assumption*. For those cases where product failures are neither exponentially distributed nor such that the single-failure assumption applies, M(t) can be approximated or solved numerically. We will demonstrate these various cases through the following examples.

4.2.1.1 FRW Policy

Example 4.4

A vacuum cleaner is produced at a cost of $50 and sold with a complete warranty on all parts and labor for a period of w years. The average cost to repair or replace a vacuum cleaner that fails under warranty is $25, which we assume is constant. The product has a mean time to failure of 1.5 years.

a. Production units enter service and function during their useful life. The failure time distribution is then exponential with parameter $\lambda = 1/\text{MTTF} = 2/3$ and

$$f(t) = \frac{2}{3} e^{-\frac{2}{3}t}, \quad t \geq 0$$

For this case, M(t) is linear and, hence, the total expected cost from Equation (4.18) is

$$C_a(w) = c_0 + c_1 \lambda w \qquad (4.19)$$

Substituting for $c_0 = \$50$ and $c_1 = \$25$,

$$C_a(w) = 50 + 25 \left(\frac{2}{3}\right) w$$

or

$$C_a(w) = 50 \left(1 + \frac{w}{3}\right)$$

b. Products enter service with failure times distributed Erlang. Let X be distributed Erlang with pdf

$$f(t) = \frac{16}{9} t e^{-\frac{4}{3}t}, \quad t \geq 0$$

the Renewal Function for this Erlang (λ, 2) pdf is given by

$$M(t) = \frac{\lambda}{2}t - \frac{1}{4} + \frac{1}{4}e^{-2\lambda t}, \quad t > 0 \qquad (4.20)$$

(See Appendix C)
For our example here, we have $\lambda = 4/3$ and $k = 2$; therefore it follows that

$$M(w) = \frac{2}{3}w - \frac{1}{4} + \frac{1}{4}e^{-\frac{4}{3}w}, \quad w > 0$$

and substituting into Equation (4.18) results in the total expected cost given by

$$C_b(w) = 50\left(1 + \frac{w}{3}\right) - \frac{25}{4}\left(1 - e^{-\frac{4}{3}w}\right), \quad w > 0$$

c. Products enter service with failure times distributed Weibull. Let X be distributed Weibull with probability density function

$$f(t) = \lambda\beta(\lambda t)^{\beta-1}e^{-(\lambda t)^\beta}, \quad \beta > 0, \ \lambda > 0, \ t > 0$$

Since we do not have a closed form of M(t) for the case of Weibull distributed failure times, we apply the method by Ayhan et al. (1999) to approximate M(w) in Equation (4.12). The total expected cost function is then given by

$$C_c(w) = 50 + 25M(w), \quad w \geq 0$$

The total expected warranty cost, C(w), for the failure time conditions given in a, b, and c are shown graphically in Figure 4.4. For the case of exponential and Erlang distributed failure times, the MTTF is 1.5 years. The Weibull distribution results in the higher cost over the range $0 < w < 3$ years.

d. Single Failure Assumption. Under this single-failure-assumption we assume $P(N(w) > 1) \approx 0$, and the right-hand term in Equation (4.18) can be replaced by the expected cost for a failure given by Equation (4.1), Equation (4.4), and Equation (4.13) for respective FRW, PRW, and combined policies. Therefore, the expected total cost is approximated by

$$C_S(w) \approx c_0 + c_1 F(w) \qquad (4.21)$$

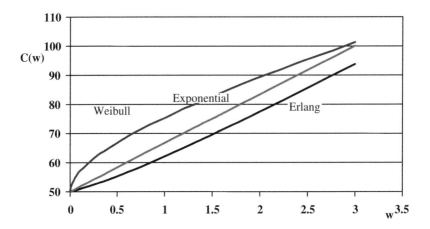

FIGURE 4.4
Total expected warranty cost vs. w.

Applying this single-failure-assumption in our vacuum cleaner example for exponential, gamma, and Weibull failure time distributions, we substitute the respective distribution function for $C(w) = 50 + 25\,M(w)$. These costs for $w = 1, 2,$ and 3 years are summarized in Table 4.1. Clearly, $C_S(w) \leq C(w)$, since $F(w) \leq M(w)$ for all $w>0$.

Upon examining the graphs in Figure 4.4 and the values for $C_S(w)$ in Table 4.1, it is clear that for these conditions the single-failure assumption does not give good approximations for the range $0 < w < 3$ years with a MTTF of 1.5 years. Note that for the number of failures during the warranty

TABLE 4.1

Expected Warranty Costs for Example 4.3

Distribution	w	$C_S(w) = 50 + 25F(w)$	$C(w) = 50 + 25M(w)$	$F(w)$
$F(w) = 1 - e^{-\frac{2}{3}w}$	1	65.80	66.67	.487
	2	68.41	83.33	.735
$\mu = 1.5 \quad \sigma^2 = 2.25$	3	71.62	100.00	.865
$F(w) = 1 - \left(1 + \frac{4}{3}w\right)e^{-\frac{4}{3}w}$	1	59.62	62.08	.385
	2	68.65	77.50	.746
$\mu = 1.5 \quad \sigma^2 = 1.13$	3	73.01	93.85	.079
$F(w) = 1 - \exp\left[-\left(\frac{5}{6}w\right)^{\frac{1}{2}}\right]$	1	64.97	75.50	.401
	2	68.12	89.25	.275
$\mu = 2.4 \quad \sigma^2 = 21.6$	3	69.86	101.25	.206

Economic Models for Product Warranties

period to be no more than 1, the time to the first failure T_1 must be less than or equal to w. For the case of T being distributed exponential with w = 1 year,

$$P(N(w) \leq 1) = P(N(w) = 0) + P(N(w) = 1)$$

$$= \frac{(\lambda w)^0 e^{-\lambda w}}{0!} + \frac{(\lambda w)^1 e^{-\lambda w}}{1!}$$

$$= \left(1 + \tfrac{2}{3}(1)\right)e^{-\tfrac{2}{3}(1)} = 0.86$$

which is relatively large for approximating M(w) by F(w).

Example 4.5

A handheld computer can be manufactured at a cost of $100 and sold with a FRW policy. Units failing during warranty will be replaced at a cost to the manufacturer of $100. The mean time to failure is 2.5 years and failure times are distributed Erlang with cumulative distribution function (cdf)

$$F(t) = 1 - (1 + .8t)e^{-.8t}, \quad t \geq 0$$

This is a standard form of the Erlang distribution with parameters $\lambda = 0.8$ and $k = 2$; therefore, applying the single-failure assumption in Equation (4.21) for the expected cost

$$C_S(w) = 200 - 100(1 + .8w)e^{-.8w}, \quad w \geq 0$$

Substituting these parameter values into Equation (4.20) with

$$M(w) = 0.4w - 0.25 + 0.25e^{-1.6w}$$

leads to the actual total expected cost given by

$$C(w) = 75 - 40w + 25e^{-1.6w}, \quad w \geq 0$$

A comparison of these cost functions is given in Table 4.2 for warranty periods of 3, 6 and 12 months.

For this example, $C_S(w)$ provides a relatively good approximation to C(w).

TABLE 4.2

Expected Warranty Costs for Example 4.3

w	F(w)	M(w)	$C_F(w)$	C(w)
0.25	.018	.045	101.75	101.75
0.50	.062	.062	106.16	106.20
1.00	0.19	0.20	119.12	120.05

4.2.1.2 PRW and Combined Policies

For the nonrenewing PRW policy, the cost of failures is dependent upon the failure times. Consequently, the expected total cost given by Equation (4.21) can be quite cumbersome. However, it follows that for a linear pro rata rate

$$C(w) \leq c_0 + E[X_1]M(w), \quad w \geq 0 \qquad (4.22)$$

where $E[X_1]$ is given by Equation (4.4) for PRW and Equation (4.13) for combined FRW/PRW policies.

Example 4.6

A tire manufacturer produces a new model of all terrain tires at a cost of $120 per unit and sells them with a 4-year PRW policy with a linear pro rata rebate. Failure times are assumed to be distributed exponential with a MTTF of 60 months.

For the case of the PRW policy in general, we substitute the unit cost given in Equation (4.4) to obtain

$$C(w) \leq c_0 + c_0 \left[F(w) - \frac{1}{w} \int_0^w tf(t)dt \right] M(w) \qquad (4.23)$$

and for exponentially distributed failure times with probability density and cumulative distribution functions

$$f(t) = \lambda e^{-\lambda t}, \quad \text{and} \quad F(t) = 1 - e^{-\lambda t}, \quad t \geq 0$$

the expected cost is

$$C(w) \leq c_0 + c_0 \{\lambda w(1 - e^{-\lambda w}) - [1 - (1 + \lambda w)e^{-\lambda w}]\} \qquad (4.24)$$

Substituting the parameter values $\lambda = 1/60$, $w = 48$, and $c_0 = 120$, we have

$$C(48) \leq 120 + 120 \left\{ \frac{48}{60}\left(1 - e^{-\frac{48}{60}}\right) - \left[1 - \left(1 + \frac{48}{60}\right)e^{-\frac{48}{60}}\right]\right\} = \$149.92$$

4.2.2 Renewing Warranty

For renewing warranties, there is never more than one failure during a warranty period. Therefore, the expected cost for a failure given by Equation (4.1), Equation (4.4), and Equation (4.13) can be simplified to determine the total cost for respective FRW, PRW, and combined policies.

Consider a product produced at a cost of c_0 and, when a failure occurs with the warranty period (0,w), it is replaced and the replacement item is covered by a new warranty. Thomas (1983) showed that the expected cost is then given by

$$C(w) = c_0 + E\left[\sum_{i=1}^{N(w)} X_i(t)\right]$$

where N(w) is a random variable representing the number of failures that occur in (0,w) and the replacement costs $X_1(t)$, $X_2(t)$, $X_3(t)$, ... are independent and identically distributed F(t). Since N(t) is a stopping time presumed to be independent of $X_1(t)$ (see Ross, 1983), it follows that C(w) is given as in Equation (4.20)

$$C(w) = c_0 + E[X_1(w)]M(w)$$

where $E[X_1(w)]$ is given by Equation (4.1) for a FRW policy and Equation (4.4) for the PRW. It follows that N(w) is distributed geometrically with

$$P(N(w) = n) = [1 - F(w)][F(w)]^{n-1}, \quad n = 1, 2, \ldots$$

therefore,

$$M(w) = E[N(w)] = \frac{1}{1 - F(w)}$$

and, hence,

$$C(w) = c_0 + \frac{E[X_1(w)]}{1 - F(w)} \qquad (4.25)$$

Now, for the FRW policy, $E[X_1]$, given by Equation (4.2), leads to the total expected cost

$$C_F(w) = \frac{c_0}{1 - F(w)} \qquad (4.26)$$

and, for the PRW policy, substituting Equation (4.4) into Equation (4.25) and simplifying gives

$$C_P(w) = \frac{c_0 - \dfrac{c_0}{w}\displaystyle\int_0^w t f(t)dt}{1 - F(w)} \qquad (4.27)$$

Example 4.7
An electrical appliance is produced at a cost of $200 and sold with a 1-year FRW renewing policy. Failure times are distributed exponential with a MTTF of 18 months. For the exponential

$$F(t) = 1 - e^{-\lambda t}, \quad t \geq 0$$

therefore, in Equation (4.26), we have

$$C_F(w) = c_0 e^{\lambda w}, \quad w \geq 0$$

and, substituting for w = 1 year and $\lambda = 2/3$,

$$C_F(1) = 200 e^{2/3} = \$389.55$$

Example 4.8
A battery manufacturer produces a premium battery for construction equipment at a cost of $300 and sells it with a 2-year linear pro rata PRW policy. Failure times follow an exponential distribution with a MTTF of 3 years. For this model, we substitute for the exponential distribution in Equation (4.27),

$$C_P(w) = \frac{c_0 - \dfrac{c_0}{w} \displaystyle\int_0^w \lambda t e^{-\lambda t} dt}{1 - e^{-\lambda w}}$$

which simplifies to

$$C_P(w) = \frac{c_0 \lambda w - c_0[1 - (1 + \lambda w)e^{-\lambda w}]}{\lambda w (1 - e^{-\lambda w})}, \quad w \geq 0$$

Substituting for c = $300, w = 2 years, and $\lambda = 1/3$ failures per year,

$$C_P(2) = \frac{300\left(\dfrac{2}{3}\right) - 300\left[1 - \left(1 + \dfrac{2}{3}\right)e^{-\frac{2}{3}}\right]}{\dfrac{2}{3}\left(1 - e^{-\frac{2}{3}}\right)} = \$483.00$$

4.3 Determining Optimum Warranty Periods

Customers value product warranty relative to the perceived quality of the product. If the quality of an item is thought to be high, they will not require as much protection or assurance against failures. For items of lesser or

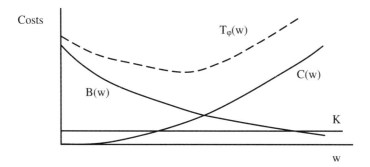

FIGURE 4.5
Expected total warranty cost $T(w) = C(w) + B(w) + K$.

unknown quality, the manufacturer must provide greater coverage. This coverage can be in the form of the amount of compensation for failures and the length of the warranty period. To determine the policy requires knowledge of failure characteristics of the product, the associated costs for replacement or repair, and some measure of the benefit to the manufacturer as a function of the warranty.

Let $B(w)$, $w \geq 0$ represent the expected cost benefit of a warranty over $(0,w)$, and $C(w)$ the expected cost of the warranty. The expected total cost is given by

$$T(w) = C(w) + B(w) + K$$

where K is the fixed administrative costs for the program. Figure 4.5 shows the typical pattern for these costs. $C(w)$ is an increasing function in $w \geq 0$, whereas $B(w)$ will decrease as the warranty interval increases. Assume both are monotonic and $T(w)$ is convex. The optimum length of warranty, w^*, therefore can be determined from simple methods. For the models that follow, we will consider an FRW policy and a quadratic benefit function with

$$B(w) = (b_0 - b_1 w)^2, \ 0 \leq b_0/b_1 \leq w \tag{4.28}$$

The total expected cost is then given by

$$T(w) = (c_0 + K) + c_1 M(w) + (b_0 - b_1 w)^2, \ 0 \leq b_0/b_1 \leq w \tag{4.29}$$

To find the optimum w, then,

$$\frac{d}{dw} T(w) = c_1 m(w) + 2(b_0 - b_1 w)(-b_1) = 0$$

and it follows that the optimum choice of w must satisfy the equation

$$c_1 m(w) + 2b_1^2 w - 2b_0 b_1 = 0 \tag{4.30}$$

where $m(w) = d/dw \, M(w)$ is the renewal density function.

4.3.1 Exponential Failure Times

When failure times are distributed exponential, as is the case during the useful life of a product, the number of failures during warranty, N(w), is Poisson and $M(w) = \lambda w$. Thus, $m(w) = \lambda$ and from Equation (4.30),

$$w^* = \frac{2b_0 b_1 - c_1 \lambda}{2b_1^2} \quad (4.31)$$

Example 4.9

Consider the example from Thomas (1999) where a nonreparable item costs $1000. With failures during useful life at a rate of 0.5 per year, each cost the manufacturer $1000. Without the warranty, the manufacturer estimates that it would be necessary to incur a cost of $2500 to market the item and, with warranty, the marketing costs would decline as $B(w) = (50-10w)^2$. Thus, $c_1 = 1,000$; $\lambda = 5$; $b_0 = 50$; $b_1 = 10$ in (4.31) and

$$w^* = \frac{2(50)(10) - (1,000)(0.5)}{2(10)^2} = 2.5 \text{ years}$$

4.3.2 Weibull Failure Times

For the failure times T distributed other than exponential M(w) and m(w) in Equation (4.29) and Equation (4.30) can be cumbersome. They can be dealt with computationally. Let T be distributed Weibull with distribution

$$F(t) = 1 - \exp[-\lambda t^\beta], \quad \lambda > 0, \quad \beta > 0$$

Tables for the associated M(t) over a wide range of values for λ and β are given in Blischke and Murthy (1993).

Suppose in Example 4.8, we have $\lambda = 0.5$ and $\beta = 2$. Substituting the parameter values into Equation (4.23),

$$T(w) = (1,000 + K) + 1,000 M(w) + (50 - 10w)^2$$

which is evaluated numerically with the results provided in Table 4.3, resulting in $w^* = 2.5$ years.

Alternatively, for given parameters ($\lambda = 1$, $\beta = 2$), M(w) can be estimated by fitting an equation through the tabled Renewal Function values. Here, the linear fit

$$\hat{M}(w) = \frac{w}{2} - 0.2$$

Economic Models for Product Warranties

TABLE 4.3
Total Warranty Costs for Weibull ($\lambda = 1$, $\beta = 2$) Failure Times Example

w	M(w)^1	1000 M(w)	(50–10w)^2	T(w)-K
0	0	0	2,500	3,500
.5	.0921	92.1	2,025	3,117.1
1	.284	284	1,600	2,884
1.5	.5127	512.7	1,225	2,737.7
2	.7549	754.9	900	2,654.0
2.25	.8789	878.9	756.25	2,635.15
*2.5	1.002	1,002	625	2,627
2.75	1.1264	1,126.4	506.25	2,632.65
3	1.2510	1,251	400	2,651
3.5	1.5007	1,500.7	225	2,725.7
4	1.7507	1,750.7	100	2,850.7
4.5	2.00007	2,000.7	25	3,025.7
5	2.2508	2,250.8	0	3,250.8

provides an adequate estimate of M(w) from which we get m(w) = $1/2$. Substituting these values into Equation (4.25), we find

$$w^* = \frac{2b_0 b_1 - 1,000(1/2)}{2b_1^2} = 2.5 \text{ years}$$

4.4 Estimating Warranty Reserve

Warranty reserves are funds that are set aside to cover forthcoming warranty expenses for a given product. Menke (1969) proposed a method, which was later generalized by Amato and Anderson (1976) and Thomas (1989), for predicting the reserve as a fraction of total sales of the product. We assume: (1) a PRW policy over (0,w), (2) items are produced and sold in lot size K at price c_1, $/unit, and (3) the failure time probability density function g(t) is known.

From Equation (4.4), the unit warranty cost is

$$X(t) = \begin{cases} c_1(1 - t/w), & 0 < t < w \\ 0, & t \geq w \end{cases}$$

Let θ be the rate of return and π the expected price change per period. The future funding requirement to cover expenses during warranty is given by

$$d(R) = X(t)d(K \cdot G(t))$$

or

$$d(R) = c_1(1-t/w)(1+\theta)^{-t}(1+\pi)^{-t} \quad (4.32)$$

Given that

$$(1+\theta)^{-t}(1+\pi)^{-t} \approx (1+\theta+\pi)^{-t} = b^{-t}$$

it follows that

$$R(w) = \left(\frac{Kc_1}{w}\right)\int_0^w (w-t)\, b^{-t}\, g(t)\, dt \quad (4.33)$$

4.4.1 Exponential Failure Times

Let the failure time T be distributed exponential with probability density function

$$g(t) = \lambda e^{-\lambda t}, \quad \lambda > 0, \quad t \geq 0$$

substituting into Equation (4.33),

$$R(w) = \left(\frac{Kc_1}{w}\right)\int_0^w (w-t)\, b^{-t}\, \lambda e^{-\lambda t}\, dt$$

$$= Kc_1\lambda \int_0^w e^{-t(\ln b + \lambda)}\, dt - \frac{Kc_1\lambda}{w}\int_0^w t e^{-t(\ln b + \lambda)}\, dt \quad (4.34)$$

Starting with the first term,

$$Kc_1\lambda \int_0^w e^{-t(\ln b + \lambda)}\, dt = Kc_1\lambda \cdot \left.\frac{e^{-t(\ln b + \lambda)}}{-(\ln b + \lambda)}\right|_0^w$$

$$= \frac{Kc_1\lambda}{\ln b + \lambda}(1 - e^{-w(\ln b + \lambda)}) = \frac{Kc_1\lambda}{\ln b + \lambda}(1 - b^{-w}e^{-\lambda w}) \quad (4.35)$$

For the second term, integrating by parts, we let u = t and

$$dv = e^{-t(\ln b + \lambda)}$$

Economic Models for Product Warranties

therefore, $du = dt$ and

$$v = \frac{e^{-t(\ln b + \lambda)}}{-(\ln b + \lambda)}$$

It follows that

$$\frac{Kc_1\lambda}{w}\int_0^w t e^{-t(\ln b + \lambda)} dt = \frac{Kc_1\lambda}{w(\ln b + \lambda)}\left[\frac{1 - b^{-w}e^{-\lambda w}}{\ln b + \lambda} - wb^{-w}e^{-\lambda w}\right] \quad (4.36)$$

Now, substituting Equation (4.35) and Equation (4.36) into Equation (4.34), it follows that

$$R(w) = \frac{Kc_1\lambda}{\ln b + \lambda}\left[1 - \frac{1 - b^{-w}e^{-\lambda w}}{w(\ln b + \lambda)}\right] \quad (4.37)$$

Example 4.10

A product is sold under a 1-year PRW policy and has failure times that are exponentially distributed with a MTTF of 2.5 years. The interest is assumed to be 5% and price changes average 4% per period.

Here, we have $\lambda = 1/2.5 = 0.4$ failures/year, $\theta = 0.05$, $\pi = 0.04$, and $w = 1$ year. Thus, $b = 1 + 0.05 + 0.04 = 1.09$ and substituting into Equation (4.37),

$$R(1) = \left[\frac{Kc_1(0.4)}{\ln(1.09) + 0.4}\right]\left[1 - \frac{1 - (1.09)^{-1}e^{-0.4(1)}}{(1)(\ln(1.09) + 0.4)}\right] = 0.3669\, Kc_1$$

or

$$\frac{R(1)}{Kc_1} = 0.3669$$

This means that for each unit sold, 36.69% of the unit cost should be set aside to cover warranty expense.

4.4.2 Weibull Failure Times

Consider the case where the failure time T is Weibull distributed with the probability density function

$$g(t) = \lambda\beta t^{\beta - 1}\exp(-\lambda t^\beta), \quad \beta > 0,\ \lambda > 0,\ t \geq 0$$

For this case, R(w) cannot be solved simply, but it can be evaluated numerically. Therefore, it would be convenient to use the following equivalent form of Equation (4.33) (see Thomas, 1989). For $\lambda > 0$ and $\beta > 1$,

$$R(w) = \frac{Kc_1\lambda\beta}{w}\int_0^w b^{-w}\int_0^t x^{\beta - 1}\exp(-\lambda x^\beta) dx\, dt, \quad x \geq 0,\ t \geq 0 \quad (4.38)$$

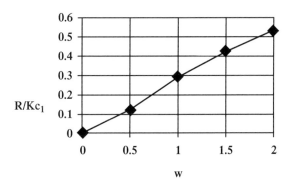

FIGURE 4.6
Fraction R/Kc_1 of unit warranty reserve cost to unit product price.

For a failure time, mean and variance of $\mu = 0.9$ and $\sigma^2 = 0.37$ (Thomas, 1999) computed the $R(w)/Kc_1$ values shown in Figure 4.6.

4.5 Exercises

1. The manufacturer of an electrical appliance has developed a new iron that has a MTTF of 30 months. Based on the FMECA, the cost of repair is estimated to be $25. Assume the iron is functioning during its useful life. Find the length of warranty period and average unit cost that will ensure the reliability during warranty is at least 0.95.

2. Assume warranty claims are made only at failure times and each cost the manufacturer c_1 units. Develop expressions for the mean unit cost for each of the following failure time distributions that are under an FRW policy with
   a. Exponential (λ)
   b. Erlang (r, λ)
   c. Weibull (β, θ)

3. Consider a product with a unit cost to the manufacturer of c_1 dollars that is sold under a FRW policy for a period of length $w > 0$. Determine the expected unit warranty costs for each of the following failure time distributions.
   a. Uniform over (a,b)
   b. Normal with mean μ and variance σ^2
   c. In Problem 2, for the case of Erlang distributed failure times let $c_1 = \$500$, $w = 3$ years, $r = 3$, and the MTTF is 4.5 years. Compute the expected unit warranty costs.

Economic Models for Product Warranties

4. Let the failure time T be distributed exponential with parameter $\lambda > 0$. Determine the variance of the unit warranty cost for an FRW policy with unit manufacturing cost c_1.

5. A product sold under an FRW policy with manufacturing cost c_1 has the failure time probability density function

$$f(t) = \lambda_1 \gamma e^{-\lambda_1 t} + \lambda_2 (1-\gamma) e^{-\lambda_2 t}, \quad t \geq 0$$

Determine the expected unit warranty cost.

6. A wide screen digital television is sold for $2000 with a 1-year warranty on all parts and services. The mean time to failure is 3.5 years. The average cost of repair is $35, regardless of whether the manufacturer or customer incurs the cost. Customers are given an option to purchase an extended additional year of warranty for $50.

   a. Given that failures are distributed exponential, determine if it is cost effective for a customer to purchase the additional warranty.

   b. At what cost would a customer be indifferent to the purchase of an extended warranty?

7. A lawn mower manufacturer produces a popular model of mower that has a mean time to failure of 2.5 years. Warranty costs average $50 per claim to the manufacturer. A 1-year FRW policy is provided with the purchase.

   a. The company has an overall annual goal to reduce costs by 20%. What is the target failure rate for this goal?

   b. How much reduction in the failure rate is necessary to make a 3-year policy cost effective for the manufacturer?

8. Compute the expected discounted unit warranty cost in problem 3a and 3b using a discount factor of 0.05.

9. Verify the result of Equation (4.10).

10. Show that the expected discounted unit warranty cost for the combined FRW/PRW policy is given by

$$E[X(t)] = \frac{c}{w_2 - w_1} \int_{w_1}^{w_2} \int_0^t e^{-\delta \tau} f(\tau) d\tau \, dt$$

11. Apply the result from Problem 10 to determine the discounted unit warranty cost for the case of exponential distributed failure times.

12. Derive an economically equivalent FRW policy for a combined FRW/PRW policy with expected cost given by Equation (4.7) when the failure times are distributed Weibull with parameters $\beta \geq 1$ and $\theta > 0$.

13. Consider a product with the failure time probability density function in Exercise 5. Show by applying Equation (2.52) in Chapter 2 that the Renewal Function is

$$M(t) = \left(\frac{\lambda_1 \lambda_2}{(1-\gamma)\lambda_1 + \gamma\lambda_2}\right)\frac{t^2}{2} + \left(\frac{\gamma\lambda_1 + (1-\gamma)\lambda_2}{(1-\gamma)\lambda_1 + \gamma\lambda_2}\right)t, \quad t \geq 0$$

Determine the optimal warranty policy for the case of a quadratic benefit function given by Equation (4.20).

14. The manufacturer of a new model of residential window is being introduced into the market. Based on preliminary testing, the mean and standard deviation of the failure time distribution is 8 and 6 years, respectively. The warranty benefit function is given by

$$B(w) = (25 - 2w)^2, \quad 0 \leq 12.5 \leq w.$$

Assume that failures occur as renewals, i.e., the times are independent and identically distributed. Find an expression for the optimal warranty using the asymptotic approximation for M(t) given in Equation (2.58) in Chapter 2.

15. Suppose items are produced and sold in lots of size K at a unit price c_0 \$/unit with each item covered by an FRW policy over the period (0,w). The unit warranty cost is then given by

$$X(t) = \begin{cases} c_0, & 0 < t < w \\ 0, & t \geq w \end{cases}$$

Let θ and π be the rate of return and expected price change per period, respectively, as in Section 4.4. Show that the warranty reserve for the future funding requirement given in Equation (4.33) can also be expressed as

$$R(w) = \frac{Kc_1}{w}\int_0^w \int_0^t b^{-u} g(u) du dt, \quad w \geq 0$$

References

Amato, H.N. and E.E. Anderson, 1976, Determination of warranty reserves: an extension, *Manage. Sci.*, 22, No. 12: 1391–1394.

Ayhan, H.J., J. Limòn-Robles, and M.A. Wortman, 1999, An approach for computing tight numerical bounds on renewal functions, *IEEE Trans. on Reliability*, 48, No. 2, pp. 182–188.

Blischke, W.R. and D.N.P. Murthy, 1993, *Warranty Cost Analysis*, Marcel Dekker, New York.

Menke, W.W., 1969, Determination of warranty reserve, *Manage. Sci.*, 15 No. 4 : B542-549.

Ross, S.M., 1983, *Stochastic Processes*, John Wiley & Sons, New York.

Thomas, M.U., 1999, Some economic decision problems in warranty planning, *Engr. Economist*, 44, No. 2, pp. 184–196.

Thomas, M.U., 1989, A prediction model for manufacturer warranty reserves, *Manage. Sci.*, 35, 12: 1515–1519.

Thomas, M.U., 1983, Optimum warranty policies for nonreparable items, *IEEE Trans. Reliabil.*, R-32, 3: 282–288.

Thomas, M.U., 1981, Warranty planning and evaluation, *IIE Annual Conference Proceedings*, pp. 478–483.

5
Product Quality Monitoring and Feedback

In Chapter 3 we defined product quality in terms of six dimensions that relate to the performance in the way an item is developed, produced, and accepted by consumers relative to their expectations. This broad interpretation of quality extends over all of the processes for developing and producing products and includes the profile of performance throughout the product life. Chapter 4 discussed the basic warranties that are applied to consumer products and models for dealing with related economic decisions. Warranty claims represent quality problems and, therefore serve an important function of providing feedback to manufacturers on product quality. All of these quality dimensions affect the number and cost of warranty claims for a product. In this chapter we present a framework for measuring and tracking overall quality performance. To maintain high quality it is important to understand the causes of quality problems, be able to track performance over time, and have some capability to provide responsive actions for improvements.

5.1 Quality System Monitoring and Control

Let us start by clarifying the meaning of the terms that pertain to manufacturing and production. Following the definitions from Groover (2001), a manufacturing enterprise is the total organization and resources that are integrated to produce products, which includes the financial and business planning functions, marketing and service, and design and engineering, along with the production and manufacturing systems. A manufacturing system is the collection of equipment and human resources for performing operations for producing products, whereas a production system is the more inclusive collection of the manufacturing operations in concert with the organization, materials, and processes, which includes the scheduling, routing, inspections, and inventory management activities that support the manufacturing system. The terms *production* and *manufacturing systems* are, however, often used interchangeably and, when reference is made to "the manufacturer,"

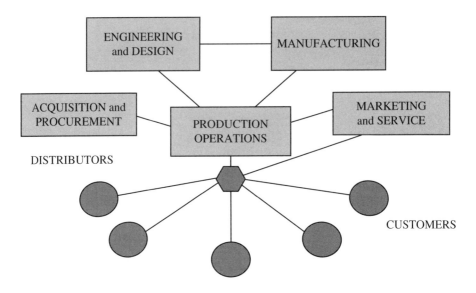

FIGURE 5.1
Primary functions in a manufacturing enterprise.

this usually refers to the manufacturing enterprise. Figure 5.1 is a diagram of a manufacturing enterprise explaining the primary functions.

During the initial stage of a product development, the item is designed, evaluated for cost feasibility, and then modeled or prototyped and tested for the design concept, which includes the materials selected for the product. The selections of the materials and the processes for making the product have to be weighed against the costs and manufacturability of the units. Extensive failure testing can be expensive, but is necessary to ensure that the failure characteristics of the product are known before products enter the market. Knowledge of the failure characteristics is also essential in developing the warranty policy for products. The effectiveness in carrying out the design and engineering functions are critical to achieve acceptable reliability and durability quality goals. The manufacturing and production operations, on the other hand, relate more to how well the processes and workmanship meet conformance quality requirements that are established through the standards for tolerances, operations, and specifications for production.

These quality problems tend to surface early in the product life, often shortly after the product enters the market. Reliability problems generally make their appearance later in the product life after the product has been used over a range of conditions. Supply chain management is part of the production system and can influence both conformance quality and reliability. Producers of assembled products, who rely heavily on outsourcing, have to maintain proper controls to ensure that the flow of materials, supplies, and subassemblies comply with the standards and requirements for product design and production. The Marketing and Service function in Figure 5.1 provides the management and direction of the sales and distribution service that includes warranty service and customer relations for promoting and marketing the products. It is through this function of the manufacturing enterprise that customers interface with the product and manufacturer. Dealers and retailers of products inadvertently serve as manufacturing representatives and often play an important role in developing and maintaining customer relations and portrayal of the perceived quality of products to consumers.

To achieve high quality products, manufacturers strive for perfection in all of the functions in Figure 5.1, and they implement strategic plans for making improvements in order to remain competitive in producing consumer products. To improve product quality requires feedback information on the quality relationships between the engineering, manufacturing, and production functions that relate to how the products are developed, produced, and used by customers.

A large manufacturing enterprise with several production, manufacturing support functions, and supply chains can have embedded within it numerous control systems of various types among the subsystems and components for maintaining standards and desired performance levels. These could be continuous controllers for maintaining control of physical measures and characteristics, such as temperature, pressure, and current flow as well as discrete control of inventory, bottlenecks or queues, and number or fraction of quality malfunctions. Details on the different types of controls used in manufacturing and production systems are discussed in Groover (2001). Most continuous quality improvement programs are based on a conceptual view of an overall total quality system that can be viewed as a discrete control system.

A total quality control (TQC) system is defined as a set of processes and procedures for achieving and maintaining high quality throughout a product's life cycle We consider the process of producing a product as a closed loop control system. Voland (1986) used a closed-loop system to describe the time varying effects of process lag and feedback on a production system. Thomas (1998) applied this concept, incorporating a double feedback loop to include conformance quality monitoring during the production cycle, followed by a post-sales customer feedback loop. The diagram for this system is shown in Figure 5.2. The materials, energy, and human resources that go into producing the item are the *inputs*. For our purposes here, the controller element represents all of the design, engineering, and preproduction processes that make up what we will loosely call "management." The "process" element represents the actual production processes that include operations, scheduling, routing, and material handling functions that are employed in its production. The first feedback loop is the information received through material and process inspections, audits, and process tracking by conventional statistical process control procedures for conformance quality.

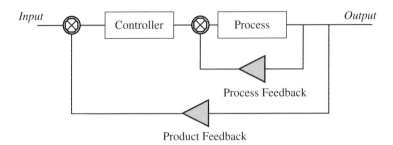

FIGURE 5.2
A production control model with warranty.

The second feedback element in Figure 5.2 represents the "product" feedback. This information is derived from consumer surveys, customer complaints, and warranty claims. In control theory terms, this loop adds stability and control sensitivity provided the feedback data are received with sufficient timeliness and accuracy to allow the controller to be responsive and make adaptive changes.

A product is generally comprised of subsystems, each made up of components and elements. Problems at the element and component level give rise to subsystem and system malfunctions. An automobile, for example, might consist of the major subsystems: (1) body, (2) engine, (3) power train, and (4) interior subsystems. Each of these subsystems has sub-subsystems and parts that are produced at various plants and supplied by a large number of vendors. So, in reality, the model in Figure 5.2 is an overall representation of several such models for the various subsystems and elements of the overall system. Therefore, to maintain control of the product quality it is necessary to impose relatively tight controls throughout the system.

5.1.1 Features for an Effective TQC System

A TQC is a system that provides a manufacturer with the necessary information to determine overall quality and make diagnostic decisions that will accomplish continuous improvement of the product and processes. To accomplish this effectively, it is necessary to provide a means of assessing product quality, an effective control capability that includes controls throughout all aspects of production and distributions, and a mechanism for translating the measures of quality into the product elements to identify strategies for making quality improvements. An effective TQC system will have the following characteristics.

1. *Complete characterization of product quality.* All quality elements should be well defined and understood before the product is introduced into the market. This includes all modes of failure, failure rates, and system dependencies. Ideally, when an item enters the

market, in addition to having known failure characteristics, the processes involved in producing it can be controlled and the only failures that occur are those due to pure chance and with no systematic errors due to the actual production. Under these conditions the product enters the market functioning in its useful life period. The failure rate of each subsystem, therefore, is constant, and the amount of failures that occur during a fixed time period (0,t) are Poisson distributed with a mean, i.e., the renewal function, that is a simple linear function of the interval.

2. *A product system configuration that will achieve quality accountability.* As mentioned above, the production control model shown in Figure 5.2 is actually a representation of several such models for the various subsystems and elements of the overall system. In designing the product system for TQC purposes, it is imperative that the system configuration is constructed such that each subsystem and major component can be treated as separate entities. In some cases the subsystem is produced by a different organization within a company or a subcontractor or vendor who could even be outside the country. The feedback information that is received for a product must be assignable to the source that produced the subsystem or element. Failures, in particular, need to be readily identified as to the location and cause within a system, disallowing any confusion with design vs. manufacturing and vendor vs. inhouse-related problems.

3. *Effective reporting conditions.* The information needed to manage and control quality must be accurate, readily available, and timely so that responsive actions can be taken.With the many data acquisition sources available today, more often than not the challenge is in getting the right information to respond in a timely fashion. This has been a major drawback with warranty data. The data used by most companies are extracted through service reports and they are not received early enough to make rapid adjustments and changes. For an effective TQC system, the data acquisition system should be developed for quality specific purposes.

4. *Feedback and diagnostic capabilities.* The feedback loops represented in Figure 5.2 should provide the necessary information to allow for the identification of the actual problems that cause failure or malfunctions. This means that the information provided in the process loop is sufficient for ensuring that the outgoing conformance quality of items is within the required standards and specifications. The product feedback loop also must allow for timely identification of the problems, which will allow for rectification and/or design changes in forthcoming models or productions.

5. *Criteria for quality improvement decisions.* Improvement decisions should be made at the system level, taking into account the impact across all dimensions of quality control. Care should be taken to

avoid suboptimal decisions that might improve some elements of quality, but only at the expense of other elements. This could lead to lower overall quality, depending on the relative weight of the various dimensions. For example, a company that provides the seats for the interior subsystem of an automobile might propose a slight change in the design of the seat adjustment control. This change might have higher component reliability with large cost savings through greater efficiency in assembly and fewer defects reported through warranties. However; if this proposed new version of seat control is not as acceptable to customers due to, say, added difficulty in its operation or it offers fewer degrees of freedom for the seat control, then it could have a serious impact on warranty costs and lost sales due to disenchanted customers.

Most producers today understand the need for incorporating customer behavior in the design of products and the importance of acceptance of these products by consumers. Conventional planning for TQC involves establishing a set of goals, developing quantitative measures for quality, and tracking the progress through feedback information, as shown in Figure 5.2. Product and process modifications are then incorporated to improve quality over time.

5.2 Multiattribute Quality Assessment

In concept, the quality of a product can be assessed quantitatively by the six attribute dimensions, which were discussed in Chapter 3. Let $Q = (Q_1,...,Q_6)$ be a vector representing the quality with the elements $Q_1 \equiv$ performance, $Q_2 \equiv$ durability, $Q_3 \equiv$ conformance, $Q_4 \equiv$ reliability, $Q_5 \equiv$ aesthetics, and $Q_6 \equiv$ perceived quality. Suppose that for each of these elements the performance can be assessed by $\pi_j(Q_j)$ that is suitably defined on a sample space Ω_j for $j = 1,...,6$. The overall product quality is then given by

$$\pi(Q) = g[\pi_1(Q_1),...,\pi_6(Q_6)] \tag{5.1}$$

defined on $\Omega = \Omega_1 \otimes \cdots \otimes \Omega_6$. Finding suitable functions $\pi_j(.)$ and $g(.)$ is generally quite difficult due to the complexities in dealing with subjective elements. Elements Q_1 through Q_4 can generally be expressed quantitatively using standard testing and measurement methods, but Q_5 and Q_6 are very subjective and generally have to be assessed through clinical classification techniques. Then to find g requires a multivariate analysis to establish and assign values of relative importance among the quality dimensions for a given product. To illustrate, for products like jewelry that do not function operationally, element Q_1 will be less important than Q_5 and Q_6, which can have extremely important dimensions for establishing the value and worth of items like necklaces, rings, and bracelets. Wristwatches, however, are considered jewelry, but

Product Quality Monitoring and Feedback

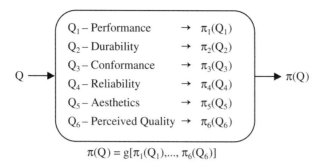

FIGURE 5.3
Measure of product quality performance.

also have functional requirements for providing the time, date, and other information, so all six of the dimensions can be significant. For some products additional elements beyond the six defined in this book might be necessary to adequately define Q. This could even further complicate the problem of reducing the multiple quality dimensions onto a simple quantitative scale for making performance and quality improvement decisions.

5.2.1 An Approach for Assessing Overall Product Quality

Assessing overall product quality is illustrated by the diagram in Figure 5.3. Each attribute Q_i is assessed by a measurement $\pi_i(Q_i)$ that assigns values onto a scale that will reflect the quality performance for each $= 1,\ldots,n$. Suppose, for example, that the product is a 140-horsepower engine for a small automobile or pickup truck. Q_1 might then be the actual power output of the engine in horsepower or brake horsepower units, the fuel consumption in miles per gallon, or the time to accelerate to a speed of 60 miles per hour. Any of these measures, or some combination of two or more, could be used for $\pi_1(Q_1)$. However, in order for the measurement process to be credible the scale should possess ordinal, transitivity, and consistency properties. Let us consider dimension Q_j that has outcome measurement values $\pi_j(Q_j) = x_j$ in Ω_j for $j = 1,\ldots,n$ and $\pi(x_1,\ldots,x_n)$ with outcomes $q \in \Omega$, then

(1) for $q^A = h(x_1^A,\ldots,x_n^A)$ and $q^B = h(x_1^B,\ldots,x_n^B)$

$$q^A > q^B \Rightarrow (x_1^A,\ldots,x_n^A) \succ (x_1^B,\ldots,x_n^B)$$

where "$\succ$" denotes "is preferred to," and
(2) for $q^C = h(x_1^C,\ldots,x_n^C)$,

$$q^A > q^B \text{ and } q^B > q^C \Rightarrow (x_1^A,\ldots,x_n^A) \succ (x_1^C,\ldots,x_n^C)$$

Consistency can be difficult to establish and maintain in assessing subjective elements, but it is important for dealing with Q_5 and Q_6.

Although it can be difficult in practice to find functions g(.) and in some cases $\pi_j(.)$ for quantifying product quality, a reasonable approach is to construct a set of weighting factors that can be used to approximate g(.) for fixed levels of measurements regarding selected elements of Q. The following procedure by Thomas (1997) consists of developing appropriate measures for all elements used to describe the product performance, combining them into a multivariate function and then constructing a quantitative scale for scoring overall quality performance.

Let $x = (x_1, ..., x_n)$ be outcome values for Q, i.e., $x_j = g_j(Q_j)$ and each has a "least" and "most" preferred level u_j and v_j, respectively for $j = 1,...,n$. Assume a linear apportionment relationship and denote the normalized values by

$$q_j = \frac{x_j - u_j}{v_j - u_j}, \quad j = 1,...,n \qquad (5.2)$$

with $q = (q_1, ..., q_n)$ being the overall measure for Q. Now, to determine an overall quality index for $\pi(Q)$ in (5.1), we find a set of weights a_j, $j = 1,...,n$ that reflects the relative importance of each of the attribute elements. Thus,

$$\pi(q) = \sum_{j=1}^{n} a_j q_j \qquad (5.3)$$

The method is summarized as follows:

1. Identify those dimensions necessary to characterize the quality of the product.
2. Assign variables for each element.
3. Establish the set of most critical variables that influence quality.
4. Construct a scale, through an iterative process of making multiple comparisons among the variables.

It is generally easier, though not essential, to divide the variables into two sets, those that are quantitative and those that are strictly qualitative. Where it is necessary to make subjective judgments in analyzing $\pi_j(.)$ and $g(.)$, as in assigning the a_j, the decision makers are assumed to be rational in the classical sense*. Therefore, individuals are capable of discriminating among alternative combinations of the quality attribute levels and establishing preferential orderings. They are also assumed to be consistent in making choices that comply with (1.) and (2.) above.

Example 5.1
An electrical appliance manufacturer produces a toaster that is comprised of a heating element, rheostat control, switch, electrical cable and plug, and a case.

---

* A rational decision maker is one who complies with the axioms for maximizing expected utility (see Marschak, 1950).

Product Quality Monitoring and Feedback

The most common failures that occur during the first month of service are due to poor soldering connections. Switches and rheostat controls are provided by a vendor and are generally of high quality, but are occasionally the source of early failure. The heating element is the critical component and once it fails the toaster is typically discarded. The toasters are tested at various stages of the assembly and the fraction of nonconforming units is 0.8. From reliability testing, the mean time to failure (MTTF) is 32 months. The company is a subscribing participant in a consumer survey service that provides quarterly reports on customer satisfaction and attitude toward the toasters and the current score is 4.6 on a (0,6)-point scale used by the firm.

Based on the given information, the significant quality dimensions considered are conformance, reliability, and perceived quality. We assume that these are the dominating variables and, while there are other factors, their impact is assumed to be significant only as through these three variables. So, let the variables x_1, x_2, and x_3 represent the quality attributes for conformance, reliability, and perceived quality, respectively. Suppose further that the most successful manufacturers achieve at best a 0.01 fraction of nonconformance in producing toasters, and some of the less prominent producers can go as low as 0.15. In other words, there is a range of 1 to 15 bad toasters that fail to meet the standards and specifications for acceptance for sales. Similarly, the best engineered toasters can have a mean time to failure (MTTF) as high as 62 months, but this can range to as low as 10 months for some of the companies just trying to break into the business. So the bounds, $(u_j, v_j), j = 1, 2, 3$ for our variables are: $0.01 \leq x_1 \leq 0.15$, $10 \leq x_2 \leq 62$, and $0 \leq x_3 \leq 6$. These data are summarized in Table 5.1.

For the given data for this toaster, $x_1 = 0.08$, $x_2 = 32$, and $x_3 = 4.3$, which from (5.2) are normalized to

$$q_1 = \frac{0.08 - 0.15}{0.01 - 0.15} = 0.50$$

$$q_2 = \frac{32 - 10}{62 - 10} = 0.423$$

$$q_3 = \frac{4.3 - 0}{6.0 - 0} = 0.717$$

TABLE 5.1

Data for Toaster Example

Variable	Attribute Level x_j	Least Preferred u_j	Most Preferred v_j
Fraction Nonconforming	0.08	0.15	0.01
MTTF	32 months	10 months	62 months
Consumer Survey Score	4.3	0	6

it follows that the quality index for this toaster is given by

$$\pi(q) = 0.50\alpha_1 + 0.423\alpha_2 + 0.717\alpha_3$$

For the assigned weights $\alpha_1 = \alpha_2 = \alpha_3 = 1/3$, the quality index is $\pi(q) = 0.547$.

5.2.2 Relative Quality Indicators

Multiattribute models for product quality are useful in planning and developing quality improvement strategies, but they generally lack the desirable properties for diagnosing specific subsystems and components. It would be nice to be able to quantitatively relate marginal differences among the various quality dimensions to $\pi(Q)$ so that engineers and managers could diagnose and interpret problems and improvement options at subsystem levels. With the absence of an ideal assessment method, manufacturers commonly resort to using warranty statistics as indirect measures of quality. Figure 5.4 illustrates the relationship between quality and warranty. When product quality is high then the expended warranty will be relatively low, and vice versa. The warranty expenditures in Figure 5.4 could be represented by warranty costs, number of claims, or a combination of both cost and volume of claims filed by customers. When a product having low quality is released to consumers then the manufacturer will eventually realize increased expenditures for warranty. This can be through the increased cost of failures and discontented consumers, or by providing warranty policy adjustments to try to offset the lower quality until it can be rectified. The relative measures from warranty feedback data that are used as indicators of product quality performance are as follows:

- Number of customer complaints
- Number of warranty claims
- Average number of warranty defects per unit sold
- Average warranty cost per unit sold

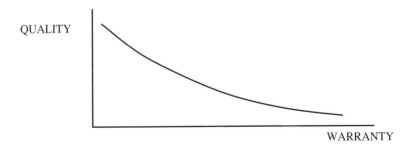

FIGURE 5.4
Product quality vs. warranty expenditures.

These are operational measures derived from tracking warranty information and often warranty claims provide the only quantitative information that is available for assessing overall product quality. The number of customer complaints is particularly important when a new product or major change is introduced. Sometimes, but not always, complaints are included in warranty claims filed by customers. Both of these sources of information are important in determining nonfailure-related problems that cause customer dissatisfaction and lower quality. Warranty claims do provide an overall assessment of the failure-related and customer difficulties in using a product, which includes problems that arise from the design, manufacture and production, and service functions. Both the number and costs of warranty claims can provide valuable quality indicators.

The number of customer complaints is particularly important when a new product or major change is introduced. Sometimes, but not always, complaints are included in warranty claims filed by customers. Both of these sources of information are important in determining nonfailure-related problems that cause customer dissatisfaction and lower quality. Warranty claims do provide an overall assessment of the failure-related and customer difficulties in using a product, which includes problems that arise from the design, manufacture and production, and service functions. Both the number and costs of warranty claims can provide valuable quality indicators.

5.3 Warranty Information Feedback Models

Warranty feedback models serve two important purposes. They provide a means for tracking the occurrence of claims and related warranty service expenses, and they provide information for diagnosing conformance quality and reliability problems. This is important for identifying strategies for implementing quality improvements.

5.3.1 The Claims Process

A descriptive model for processing warranty claims is shown in Figure 5.5. The products are released from a production system and enter the market according to an operations schedule, which we will treat as being on a monthly basis. This could be any time period, such as daily, weekly, or quarterly, but monthly planning is very common in many industries. It is also common to track the claims for products by the time period in which they were produced. Though ideally the processes for making the products will not vary significantly and they will provide constant quality over time, it is not always the case, particularly during the early production periods of a new operations schedule. The tracking of warranty information for accountability purposes, therefore, is done by the month of production (MOP) for the produced items. So, in Figure 5.5, n_{P1} units of the product are produced during MOP-1, the first month of production. This could be the

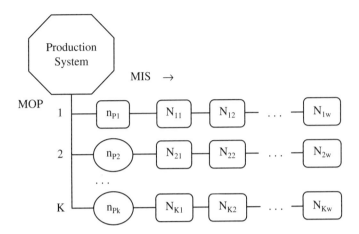

FIGURE 5.5
Descriptive model of warranty claims processing.

first month of a business year, the first month of the production of a new product, or even the first month of a calendar year, January.

For modeling purposes, the products produced during a given MOP will be treated as a lot within which all units are treated as having the same quality characteristics. In other words, an item that is produced on the first day of a MOP is the same as one produced on the last day. The units will also age by these lots and they are tracked on a monthly basis by their month in service (MIS). The claims that occur in a particular MIS are considered uniform for that month. So, for the n_{p1} units produced in MOP-1, N_{11} claims will be filed by customers during the first MIS. During MIS-2, there will be N_{12} claims which can include claims filed by customers who filed during MIS-1. Also, a given customer might file more than one claim in a given MIS. The total number of claims that will accumulate from the n_{p1} products from MOP-1 will be

$$N_1 = N_{11} + N_{12} + \cdots + N_{1w} \tag{5.4}$$

where w is the length of the warranty period in months. For each MOP-i, the n_{pi} items produced will similarly transition through w MIS periods with N_{ij} claims occurring during MIS-j, for j = 1,...,w.

5.3.2 Poisson Warranty Claims

Let N_{ij} and N_i be discrete random variables defined on $\Omega = \{0,1,...\}$ for i=1,...,K, and j = 1,...,w. The following additional assumptions will apply in developing a model for the claims events.

1. Warranty claims occur as independent events and at a constant probability of occurrence, $0 \le p_{ij} \le 1$ for a given MOP-i and within a MIS-j.

2. The probability of a single customer having more than one claim within a single MIS is negligible.
3. For a given MOP-i and MIS-j, the cost of warranty service for claim k, say X_{ijk}, is independent of the occurrence of the claim.

Consider the n_{Pi} products produced in month i, each of which will generate a warranty claim in month of service $j = i+1, i+2, ..., j+w$ with probability p_{ij}. It follows that the distribution of the number of claims in MIS-j is distributed binomial with

$$P(N_{ij} = n) = \binom{n_{Pi}}{n} p_{ij}^n (1-p_{ij})^{n_{Pi}-n}, \quad n = 0, 1, ..., n_{Pi} \tag{5.5}$$

Now let the claim probability p_{ij} grow small simultaneously as the number of products n_{Pi} grows large in such a manner that the product $n_{Pi} p_{ij}$ approaches the constant λ_{ij}. It can then be shown that the distribution of N_{ij} in (5.5) approaches the Poisson distribution with

$$P(N_{ij} = n) \simeq \frac{\lambda_{ij}^n}{n!} e^{-\lambda_{ij}}, \quad n = 0, 1, ... \tag{5.6}$$

This approximation becomes close as the number of products is large and the probability of an occurrence is small, which is generally the case for the high-volume production of consumer products.

Example 5.2

A widget manufacturer produces 10,000 widgets per month starting in January, which are then sold with a 1-year free replacement warranty (FRW) policy. The probability of a widget having a flaw or not functioning according to customer expectations varies during the age of the widget, ranging from 0.01 to 0.025 as shown in Figure 5.6.

So for those widgets produced in January, i.e., = 1, the probability of a warranty claim event during the first MIS is $p_{11} = 0.01$. This corresponds to the calendar month February, since we are starting the production cycle in January. So, the count of warranty claims N_{11} is treated as having a Poisson distribution in (5.6) with $\lambda_{11} = (10,000)(0.01) = 100$. A sample realization of outcomes of this random variable is shown in Figure 5.7.

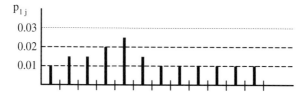

FIGURE 5.6
Probabilities of warranty claim events for MOP-1 during MIS-j.

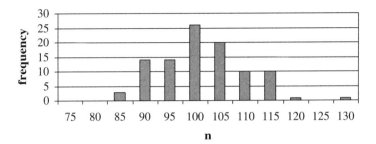

FIGURE 5.7
A sample realization of N_{11} distributed Poisson with $\lambda = 100$ widgets.

To determine the cumulative number of claims over a fixed period of time, we denote the cumulative number of claims generated from MOP-i over s months in service by the random variable

$$N_i(s) = \sum_{j=1}^{s} N_{ij} \tag{5.7}$$

It follows that for the case where $N_{i1}, N_{i2}, ..., N_{is}$ are independent and Poisson distributed with mean rates λ_{ij} for $j = 1,...,s$, then $N_i(s)$ is also distributed Poisson with the mean rate $\lambda_i = \sum_{j=1}^{s} \lambda_{ij}$. To establish this, we will apply probability transforms to (5.7). The probability generating function for the Poisson distribution in (5.6) is

$$P_{N_{ij}}(z) = e^{-\lambda_{ij}(1-z)}, \quad |z| < 1$$

Since $N_{ij}, j = 1,...,w$ are independent random variables, then the distribution of $N_i(s)$ has the probability generating function given by

$$G_{N_i(s)}(z) = \prod_{j=1}^{s} P_{N_{ij}}(z) = \prod_{j=1}^{s} e^{-\lambda_{ij}(1-z)}$$

or

$$G_{N_i(s)}(z) = e^{-(1-z)\sum_{j=1}^{s} \lambda_{ij}}, \quad |z| < 1 \tag{5.8}$$

Inversion of (5.8) leads to the Poisson distribution

$$g_{N_i(s)}(n) = \frac{\left(\sum_{j=1}^{s} \lambda_{ij}\right)^n}{n!} e^{-\sum_{j=1}^{s} \lambda_{ij}}, \quad i = 1,...,I \tag{5.9}$$

TABLE 5.2

Mean Number of Widget Warranty Claims for MOP-1

MIS-j	1	2	3	4	5	6	7	8	9	10	11	12
λ_{1j}	100	150	150	200	250	150	100	100	100	100	100	100

Example 5.3

Continuing with the widget production: since $n_{p_i} = 10,000$ for all i=1,...,12, then $\lambda_{1j} = 10,000 p_{1j}$ and the expected number of claims for widgets produced in January for each MIS are given in Table 5.2.

A sample realization of this process over a 12-month period of monthly service intervals is illustrated in Figure 5.8a,b. The outcomes of N_{1j}, which are simulated values for the number of claims by MIS for the widgets produced during MOP-1, are shown in Figure 5.8a. The corresponding cumulative number of claims, $N_1(s)$, increases as a counting function over the s periods of MIS, as shown in Figure 5.8b.

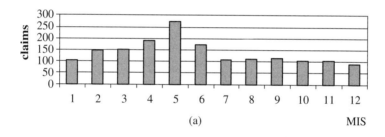

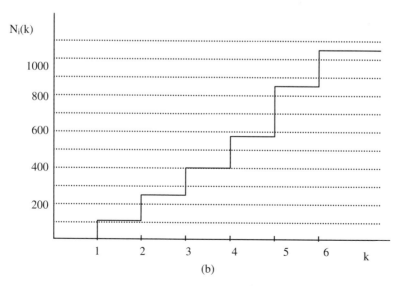

FIGURE 5.8a,b
(a) N_{1j} claims for widgets by MIS; (b) Cumulative number of claims produced in MOP-1.

So far we have considered only the warranty claims process for a single month of production in our example. This process is repeated for each production month, thus generating a large number of claims that are reported each month for what might be considered moderate production levels with relatively high quality for consumer products.

The Poisson model is common in modeling failure processes because of its inherent properties from the exponential occurring event times. This corresponds to the elements of the product functioning during their respective useful life periods with constant failure rates, as they should if the product is properly engineered and tested before it is placed on the market. Therefore, only chance events occur and generate warranty expenses.

From (5.9), the expected number of claims for MOP-i over s months follows directly,

$$E[N_i(s)] = \sum_{j=1}^{s} \lambda_{ij} \qquad (5.10)$$

To determine the total warranty cost over s periods of service, we define the random variables

$$X_{ijk} \equiv \text{the cost for claim k from MOP-i during MIS-j}$$

which for k = 1,2,... are independent and identically distributed, and by assumption (3) are independent of N_{ij}. It then follows that for items produced in MOP-i with j months in service, the expected warranty cost is given by

$$C(i,j) = E\left(\sum_{k=1}^{N_{ij}} X_{ijk}\right) = \lambda_{ij} E(X_{ijk}) \qquad (5.11)$$

The random variable $N_i(s)$ is a compound Poisson process and Equation (5.11) is known as Wald's Equation. While these results are quite intuitive, the proof of (5.11) involves arguments from renewal theory, which can be found in Ross (1983).

Example 5.4

Suppose in Example 5.3, the warranty costs for the claims in February for the claims produced in January are normally distributed with mean cost μ_{ij} and variance σ_{ij}^2. From (5.11) the expected cost is then

$$C(1,1) = \lambda_{11}\mu_{11} = 100\mu_{11}$$

So, if the average cost for a warranty claim in $50 and the variance is 25 $², then C(1,1) = $5000.

Product Quality Monitoring and Feedback

The variation about the number of claims and cost are important in interpreting the quality performance and the stability of the processes for producing the products. To examine the variation about the cost, it follows that a 2-sigma range about the warranty cost for MOP-i and MIS-j (see Appendix C) is given by

$$C(i,j) - 2\sqrt{\lambda_{ij}\left[Var(X_{ijk}) + \left(E(X_{ijk})\right)^2\right]} \leq \sum_{k=1}^{N_{ij}} X_{ijk}$$

$$\leq C(i,j) + 2\sqrt{\lambda_{ij}\left[Var(X_{ijk}) + \left(E(X_{ijk})\right)^2\right]} \quad (5.12)$$

For the case of X_{ijk}, independent and identically distributed normal with mean μ_{ij} and variance σ_{ij}^2 for each $k = 1,\ldots,K$,

$$C(i,j) - 2\sqrt{\lambda_{ij}\left(\sigma_{ij}^2 + \mu_{ij}^2\right)} \leq \sum_{k=1}^{N_{ij}} X_{ijk} \leq C(i,j) + 2\sqrt{\lambda_{ij}\left(\sigma_{ij}^2 + \mu_{ij}^2\right)} \quad (5.13)$$

For the parameter values $\mu_{11} = 50$ and $\sigma_{11}^2 = 25$ given in Example 5.4, the 2-sigma range about the warranty cost for MOP-1 and MIS-1 is

$$\$3,999 \leq \sum_{k=1}^{N_{ij}} X_{ijk} \leq \$6,001$$

5.3.3 Poisson Warranty Claims with Lags

In order for a warranty feedback system to be effective, it is essential that there is full accountability for the times of occurrence of the warranty claims. This is necessary both to relate the claims events with the appropriate occurrence times and the month that they were produced. The descriptive model of Figure 5.5 presumes that all claims are reported during the month in which they occur. There are many manufacturing enterprises for which delays exist in reporting portions of the claims generated for some period of time. There are a variety of reasons for these delays that range from work and production scheduling issues, reporting errors, organizational inefficiencies, and other circumstances. Reporting delays are certainly undesirable and should be kept to a minimum, but when they exist it is necessary to account for them in the reporting process.

To illustrate the warranty claims reporting process when reporting lags are present consider the flow of claims from units produced in the month of January for the production system shown in Figure 5.9. During February, which is the first month of service, 100 warranty claims were filed by customers,

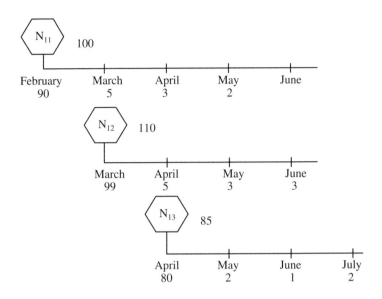

FIGURE 5.9
Lagged reporting for products produced during MIS-1.

but only 90 were actually reported in the month of February, with 5 reported in March, 3 in April, and 2 in May. During the second month of service, this same production lot had $N_{12} = 110$ claims occur in March during which 99 were reported within the month, but, for the remaining claims, 5 were reported in April, 3 in May, and 3 in June. So during MIS-3 (i.e., April), a total of 85 claims entered the report processing system consisting of 80 from MIS-3, 5 delayed from MIS-2, and 3 delayed from MIS-1.

Now, to model the warranty claims reporting process with reporting lags up to r months from the month in which they occurred, we let L be a discrete random variable representing the reporting lag in months for a claim occurring in MIS-j, and its distribution is

$$\alpha_k = P(L = k), \quad k = 0, 1, \ldots, r, \quad (5.14)$$

where $0 \leq \alpha_k \leq 1$ and $\sum_k \alpha_k = 1$. We further define the random variable $N_{ij|L}$ as the number of claims that are generated from MOP-i during MIS-j and reported with lag L. For example, in Figure 5.8 the outcome for $N_{12|L=0}$ has the outcome of 99 claims and $N_{12|L=2}$ is 3 claims. It follows that if $N_{i1}, N_{i2}, \ldots, N_{ir}$ are independent and Poisson distributed with mean rates of λ_{ij} for $j = 1, \ldots, r$, then $N_{ij|L}$ is also distributed Poisson. To show this we note that N_{ij} with lags is a filtered process that can be viewed as a flow of claims that are channeled to the processes $N_{ij|L=0}$ through $N_{ij|L=r}$ from an imaginary switch as shown in Figure 5.10. The particular direction, say k, that a claim takes occurs with a constant probability α_k and is independent of the occurrence of any particular

Product Quality Monitoring and Feedback

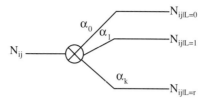

FIGURE 5.10
A filtered Poisson warranty claims reporting process with lags.

event. The distribution for the number of claims lagged by k months is found by conditioning on N_{ij} as follows,

$$P(N_{ij|L=k} = m) = \sum_{n=m}^{\infty} P(N_{ij|L=k} = m | N_{ij} = n) P(N_{ij} = n) \quad (5.15)$$

The first term in this summation is binomially distributed

$$P(N_{ij|L=k} = m | N_{ij} = n) = \binom{n}{m} \alpha_k^m (1-\alpha_k)^{n-m}, \quad m = 0,\ldots,n \quad (5.16)$$

Therefore, substituting into (5.15)

$$P(N_{ij|L=k} = m) = \sum_{n=m}^{\infty} \binom{n}{m} \alpha_k^m (1-\alpha_k)^{n-m} \frac{\lambda_{ij}^n}{n!} e^{-\lambda_{ij}}$$

$$= \sum_{n=m}^{\infty} \frac{n!}{(n-m)!m!} \alpha_k^m (1-\alpha_k)^{n-m} \frac{\lambda_{ij}^n}{n!} e^{-\lambda_{ij}}$$

$$= \frac{(\alpha_k \lambda_{ij})^m}{m!} e^{-\lambda_{ij}} \sum_{s=0}^{\infty} \frac{[(1-\alpha_k)\lambda_{ij}]^s}{s!}$$

and since the series term converges to $e^{(1-\alpha_k)\lambda_{ij}}$,

$$P(N_{ij|L=k} = m) = \frac{(\alpha_k \lambda_{ij})^m}{m!} e^{-\alpha_k \lambda_{ij}}, \quad k = 0,1,\ldots,r \quad (5.17)$$

For the case of k = 0, the claims are reported within the month in which they occur and α_0 is generally quite large relative to the other lag probabilities. The average number of claims that occur in a particular month, for a given MOP-i and MIS-j, is the average number reported without a lag, plus the expected accumulation of those reported over the next r months. It follows that

$$E(N_{ij}) = \sum_{k=0}^{r} \alpha_k E(N_{ij|L=k}) = \sum_{k=0}^{r} \alpha_k \lambda_{ij} \qquad (5.17)$$

Equation (5.17) gives the expected number of claims for a given MOP-i and MIS-j for claims with lags up to r months from their occurrence. So, on any given month t, the total number of warranty claims that are reported, i.e. are received at the central claims processing center is given by

$$NR(t) = \sum_{i=1}^{t-1} \sum_{j=1}^{t-i} N_{ij|t-i-j}, \quad t = 1, 2, \ldots \qquad (5.18)$$

Therefore, for Poisson distributed warranty claims

$$E[NR(t)] = \sum_{i=1}^{t-1} \sum_{j=1}^{t-i} \lambda_{ij} \alpha_{t-i-j} \qquad (5.19)$$

Consider the flows of warranty claims for the three production months shown in Figure 5.11. Starting with the first MOP in January, some claims will arrive at the reporting center in February from quality problems that occurred in January and have 1 MIS, while others occur in January with 1 MIS, but can be delayed possibly up to 5 months from their occurrence. The same is true for each MIS 2, 3, and similarly for the February MOP, from the $N_{2,1}$ claims that occur during February with 1 MIS reported in March without lag. But others will trickle in with the claims reported in April, May, and at various delay times of up to 5 months. In other words, $N_{2,1|0}$ claims are reported without lag, while $N_{2,2|1}$, $N_{2,2|2}$, etc. will arrive in months March through August. The pattern continues throughout the production cycles. During the month of March, the number of reported claims is

$$NR(3) = N_{1,1|1} + N_{1,2|0} + N_{2,1|0}$$

and

$$E[NR(3)] = \lambda_{1,1}\alpha_1 + \lambda_{1,2}\alpha_0 + \lambda_{2,1}\alpha_0$$

5.4 MOP/MIS Charts

In the previous section, we considered the Poisson model for warranty claims processing. The occurrences and counting of claims events were represented by random variables and the results (5.10), (5.11), and (5.17), are in terms

Product Quality Monitoring and Feedback 105

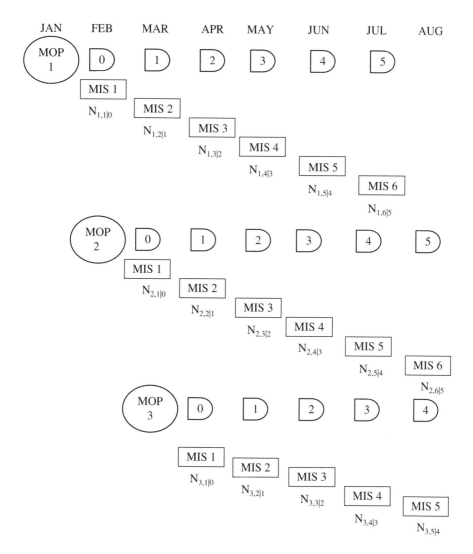

FIGURE 5.11
Warranty claims reporting with lags for 3 months of production.

of expected values. In this section, we will consider the outcomes of the random variables N_{ij} and $N_{ij|k}$, representing the number of warranty claims for MOP-i and MIS-j. The standard method for tracking the measures of relative quality of product elements from warranty claims is by each MOP and MIS. We will assume that units are sold during the same month they are produced. Let $m_{i,j} \equiv$ number of claims from items produced in MOP-i during MIS-j.

To illustrate the MOP/MIS tracking procedure and introduce the relative quality measures, we will consider the following example.

TABLE 5.3

MOP/MIS Chart of Warranty Defects (× 1000) for Golf Cart Example

MOP/Vol.		MIS											
		1	2	3	4	5	6	7	8	9	10	11	12
1 (Jan)	1000	24	18	19	22	20	19	17	18	14	20	22	16
2 (Feb)	1000	16	16	15	14	17	20	19	16	14	12	12	
3 (Mar)	2000	18	10	9	12	13	11	9	9	11	10		
4 (Apr)	2000	14	12	10	14	12	13	8	14	15			
5 (May)	3000	12	11	14	12	14	13	10	8				
6 (Jun)	3000	15	8	12	14	6	9	7					
7 (Jul)	3000	13	10	6	11	12	10						
8 (Aug)	2000	12	7	9	9	11							
9 (Sep)	2000	11	11	7	10								
10 (Oct)	1000	14	8	8									
11 (Nov)	1000	10	8										
12 (Dec)	1000	12											

Example 5.5

A manufacturer of utility vehicles produces a golf cart that is sold under a 1-year FRW policy. Units are made to order with production scheduled on a monthly basis. Warranty claims are assumed to be time-dependent due to learning effects in production. The production schedule and reported warranty claims for the past 12-month period are given in Table 5.3

The aging process of warranty claims for this example is illustrated in Figure 5.12. Items produced in January, MOP-1, will have $m_{1,1} = 24$ malfunctions

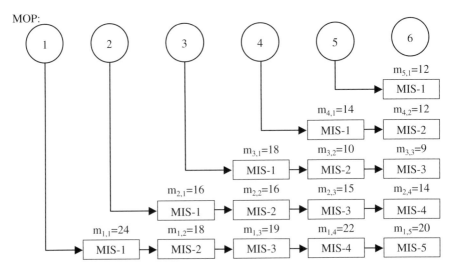

FIGURE 5.12
Aging process for MOP-1 to MOP-5 by MIS.

Product Quality Monitoring and Feedback 107

during their first month in service, $m_{1,2} = 18$ in MIS-2, $m_{1,3} = 19$ in MIS-3, and so forth.

5.4.1 Average Defects per Unit Sold (DPUS)

The matrix of m_{ij} values in Table 5.3 is called a MOP/MIS chart. During the month of January, 1000 golf carts were produced and 1 month later, in February, 24 malfunctions were claimed for the units with 1 MIS. In February of that year, another 1000 units were produced and, after 1 month, 16 defects were reported for units with 1 MIS and, for that same MOP at 2 MIS, another 16 defects were reported.

At a particular month t, we define the average number of defects per unit sold (DPUS) for items during MIS-j by

$$D_j(t) = \frac{\sum_{i=1}^{t} m_{i,j}}{\sum_{i=1}^{t} n_i}, \quad t = 1,...,w \quad (5.20)$$

So, the DPUS in May for golf carts that were in their first MIS is

$$D_1(5) = \frac{\sum_{i=1}^{5} m_{i,j}}{\sum_{i=1}^{5} n_j} = \frac{m_{1,1} + m_{2,1} + m_{3,1} + m_{4,1} + m_{5,1}}{n_1 + n_2 + n_3 + n_4 + n_5}$$

$$= \frac{24 + 16 + 18 + 14 + 12}{1000 + 1000 + 2000 + 2000 + 3000} = 9.33 \times 10^{-3}$$

The common method used to track relative quality performance is to construct MOP/MIS graphs from the warranty data. This is a plot of the running average number of defects per unit sold over time. For a given MIS-j, values of $D_j(t)$ are plotted against each MOP-t, t = 1,2,.... For the MOP/MIS chart in Table 5.3, we have for MIS-1

$$D_1(1) = \frac{24}{10} = 2.4 / 100 \text{ units sold}$$

$$D_1(2) = \frac{24 + 16}{10 + 10} = 2.0$$

...

$$D_1(12) = \frac{24 + 16 + 18 + \cdots + 12}{10 + 10 + 20 + \cdots + 10} = 0.76$$

The resulting graph is shown in Figure 5.13.

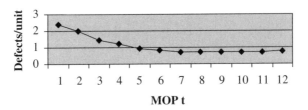

FIGURE 5.13
MOP/MIS graph for units with one MIS.

5.4.2 Average Warranty Cost per Unit Sold (CPUS)

Warranty costs are estimated and tracked by this same computational procedure, using MOP/MIS charts similar to Table 5.1 that contain marginal warranty expenses. The CPUS for MIS-j at time t is given by

$$Z_j(t) = \frac{\sum_{i=1}^{t} z_{i,j}}{\sum_{i=1}^{t} n_i}, \quad t = 1,...,w \quad (5.21)$$

where z_{ij} is the warranty cost for units produced in MOP-i that are in their MIS-j. MOP/MIS graphs are also developed for CPUS using the same plots as was used for tracking DPUS.

The DPUS and CPUS are both tracked to provide quality feedback over time. In order to use these relative warranty measures as an indicator for $\pi(Q_1,...,Q_6)$, we need to account for the effects on $Q_1,...,Q_6$. Here we have focused on the overall product, but in practice, MOP/MIS charts for average defects and cost are developed for subsystem and component levels as well. Subsystems are further broken down to the appropriate component level necessary to capture details to identify causes, effects, and diagnostic indicators that properly prescribe Q. Other breakdowns of warranty data can be useful in identifying dependencies that can occur due to patterns of usage and customer behavior. For example, customers of a given type (say age, income level, or geographic region), might generate different D(t) and Z(t) values. This type of analysis is particularly important in understanding the nonfailure related malfunctions and related causes for customer dissatisfaction.

5.4.3 MOP/MIS Charts with Lag Factors

The DPUS and CPUS measures given in (5.20) and (5.21) apply when the warranty data are free of timing or delays due to the sales and warranty

Product Quality Monitoring and Feedback

service processes. In other words, the products are produced on schedule and immediately sold and enter service, and when customers file warranty claims, the dealers or service agents immediately submit the reports to the manufacturer or producer. When time is an influencing factor in reporting warranty claims then it becomes necessary to adjust the data accordingly.

There are two types of delays that can occur with warranty data. A good example for illustrating these delays is in the manufacture of automobiles. Consider a final assembly plant that produces cars that are shipped to dealers for delivery and sale to customers. While many of the cars are produced-to-order, i.e., the customer already purchased a car and is awaiting delivery, a large volume of cars are sent to dealers who will sell the units from their lots. This includes network trading where dealers negotiate trades among themselves to fulfill customer preferences in models and options. As a result, the actual time that a vehicle is placed in service can vary from a few days to several months. This can affect the warranty data. A second type of delay occurs after cars are in service. A customer arrives at a dealer with a defective car and the dealer provides the necessary repair and parts replacement service, but does not report the problem or problems to the manufacturer until some later time. As a result of these delays, the DPUS and CPUS curves will tend to be underestimated and can lead to serious problems in projecting the eventual long-run average number of claims and warranty costs per unit sold. To more properly estimate these curves, we need to use a more detailed breakdown of the data than the monthly m_{ij} values. The absence of some of the information on the claims makes the problem one of prediction rather than parameter estimation.

5.4.3.1 Lag Factor Model

We assume that warranty claims occur as Poisson events at a rate λ_t, which is consistent with the argument that products are sold to function during their useful life and, hence, have exponential distributed times between reported defects (see Appendix C.2). Denote by $n_{x,t,r}$ the number of claims at age t that are put in service on day x with reporting lag r in days. We want to predict the long-run average number of warranty claims by age t for cars put into service over a period $(0, \tau)$ given by

$$m(\tau) = \frac{\sum_{x=0}^{\tau} \sum_{r=0}^{\infty} n_{x,t,r}}{\sum_{x=0}^{\tau} N_x} \tag{5.22}$$

and

$$M(t) = \sum_{u=0}^{t} m(u) \tag{5.23}$$

The lag factors are time adjustments for the data that can be known deterministic values from seasonal effects in marketing or other trend effects for certain products, or they could be represented by a random variable L with a discrete probability function $p_L(r) = P(L=r)$, $r = 0,1,...$, and

$$F_r = P(L \leq r) = \sum_{k=0}^{r} p_L(k) \tag{5.24}$$

Kalbfleisch et al. (1991) developed maximum likelihood estimators, λ_t, for known lag probabilities

$$\hat{\lambda}_t = \frac{\sum\sum_{x+r \leq T-t} n_{x,t,r}}{U_{T-t}} \tag{5.25}$$

where

$$U_{T-t} = \sum_{x=0}^{T-t} N_x F_{T-t-x} \tag{5.26}$$

is an adjusted count of the number of cars at risk at day t. Thus, the estimate $\hat{\lambda}_t$ in (5.25) is used to estimate $m(\tau)$ and $M(t)$ in (5.22) and (5.23). For the case of grouped data that are quite common for monthly reporting, the average number of claims per car over an interval (a,b) is estimated by

$$\hat{M}(a,b) = \frac{\sum_{t=a}^{b}\left(\sum\sum_{x+r \leq T-t} n_{x,t,r}\right)}{\frac{1}{2}\left[U_{T-a} + U_{T-b}\right]} \tag{5.27}$$

To illustrate this calculation, we consider the following example from Kalbfleish, et al.

Example 5.6

A constant flow of cars is put into service on days $x = 0,1,...,364$. The lag probabilities $p_L(r)$, $r = 0,1,...$ are known and estimates for the number of cars at risk at day t from (5.23) are given in Table 5.4.

The $U_{t-\tau}$ values given in Column 2 are factored for F_{T-t-x} from (5.26). Column 4 gives the translation of these values over the monthly time intervals (a,b) of Column 3. The estimated number of claims, $\tilde{M}(a,b)$ for given observed values $\sum_a^b n_{.t.}$, i.e., the numerator of (5.27), are computed and the predicted average number of claims per car, $\tilde{\Lambda}_t$, given in Column 7.

Warranty data can be quite variable and plagued with problems caused by misclassification of failures and problems. The reporting systems often are not designed specifically for quality and warranty feedback and, therefore, the large variations and time lags in getting information limit the application of

TABLE 5.4

Estimated Expected Number of Claims Per Car by Age 364 Days

t	$U_{t-\tau}$	(a,b)	$\frac{1}{2}\left[U_{T-a}+U_{T-b}\right]$	$\sum_{t=a}^{b} n_{t.}$	$\tilde{M}(a,b)$	$\tilde{\Lambda}_t$
0	33,550	—	—	—	—	—
30	30,550	0,30	32,050	2,064	0.06440	0.06440
60	27,450	31,60	29,000	1,726	0.05952	0.12392
90	24,550	61,90	26,000	1,586	0.06100	0.18492
121	21,350	91,121	22,950	1,416	0.06170	0.24662
151	18,450	122,151	19,900	1,132	0.05688	0.30350
181	15,350	152,181	16,900	1,053	0.06231	0.36581
211	12,450	182,211	13,900	870	0.06259	0.42840
242	9,250	212,242	10,850	648	0.05972	0.48812
272	6,352	243,272	7,801	470	0.06025	0.54837
303	3,250	273,303	4,801	294	0.06124	0.60961
333	635	304,333	1,943	136	0.07001	0.67962
364	0.83	334,364	317.9	10	0.03146	0.711075

the results to long-term improvements. The lag factors described in this section are applied to try to overcome these delays in reporting and allow useful and timely information from warranty claims.

5.5 Exercises

1. Identify the quality dimensions that best characterize each of the following products.

    a. A portable power generator designed for recreational use has a power output of 1850W and is sold with a 1-year FRW policy. Features include two handles for ease in transporting, a 12v/15 amp battery charger and charging cables, two 120v outlets and two 12v DC outlets. The generator runs up to 7 hours at 50% load.

    b. A furniture manufacturer produces a premium recliner chair that consists of a seat, back, rocker platform, foot rest, and two arm subassemblies. The platforms and footrests are produced inhouse where the units are assembled, but the seats, backs, and arms are supplied by another company. Each assembled chair is inspected and repaired as necessary before they are released to the market. The chairs have a 1% failure rate, generally from the recliner mechanism in the platform, and the MTTF is 45 months. In addition, periodically the wood used by the supplier is too soft to hold the T-nuts connecting the subassembled units firmly and they loosen during the first month of use. The chairs have consistently scored in the range of 4.2 to 4.5 on a 6-point annual consumer survey during the past 5 years.

c. Snow Fox is a lightweight 8.5 amp 110v/1100w electric snow thrower with an ergonomic double handle that allows the user to maneuver in compact areas, like stairs and landings, while an adjustable three-position discharge throws snow in desired locations. The motor is double insulated to allow for cold weather starts with a cord retention feature that reduces tangles. The unit has a 6-in. intake height and a 12-in clearing width. A limited 2-year FRW is provided that covers the motor and parts.

2. Suppose the quality of the toaster in Example 5.1 is further assessed after 1 and 2 years. For each of the following conditions, compute the quality index and interpret your result.

   a. One year later; $x_1 = 0.1$, $x_2 = 45$, and $x_3 = 3.0$.
   b. Two years later; $x_1 = 0.08$, $x_2 = 20$, and $x_3 = 3.0$.

3. Develop an alternative model to (5.3) for assessing product quality using the product form

$$\pi(q) = \prod_i q_i^\gamma$$

4. The quality of an automobile is assessed by its cost, reliability during the warranty period, and combined operational performance and appearance as determined from a customer satisfaction survey measured on a 4-point Likert scale as follows:

Q-element	Level:	1	2	3	4
q_1	Cost ($)	(<20K)	(20K–30K)	(30K–40K)	(>40K)
q_2	R(w)	(<0.9)	(0.9,0.95)	(0.95,0.99)	(>0.99)
q_3	Satisfaction	1	2	3	4

   a. Construct an approximate scale for the overall quality performance $U(q_1, q_2, q_3)$.
   b. Suggest some quality improvement options.

5. A ballpoint pen consists of a two-piece body with an ink cartridge insert. The upper body has a snap mechanism built in for engaging and dispensing the cartridge. The lower body has a built-in spring and the two pieces are thread connected. Ink cartridges are supplied by a vendor.

S_1	Upper body — male thread
S_{11}	Snap mechanism
S_2	Ink cartridge
S_3	Lower body — female thread
S_{31}	Spring

The common modes of failure for the assembled pen are (1) misalignment of the threads, (2) off-set snap mechanism, and (3) dry ink cartridge. The fraction of outgoing quality from quality control inspections and audits is estimated to be 0.96. The pens have an estimated lifetime performance of 12,000 usage hours, MTTF of 25 months, and ranks 8 out of the 12 leading pens in its class.

a. Give the quality dimensions that best characterize the pens and rank order them in relative importance.

b. Given that the competition by other manufacturers producing comparable pens will have average outgoing conformance quality of 0.01 to 0.10 defects and MTTF ranging from 18 to 30 months, construct a quality index using the additive model of (5.3).

6. Suppose the widget manufacturer in Example 5.2 produces widgets according to the following monthly schedule (widgets × 1000).

Jan	Feb	Mar	Apr	May	Jun	Jul	Aug	Sep	Oct	Nov	Dec
10	12	12	15	15	20	15	15	12	12	10	10

The warranty claim probabilities are equal for all months of production, i.e. $p_{1,j} = p_{2,j} = \ldots = p_{12,j}$, and are given in Figure 5.6.

a. Compute the mean number of claims for each MOP.

b. Compute the expected number of claims for MOP-i over 6 months and over 11 months.

7. The number of warranty claims per month for a particular product occurs as Poisson events with the mean number of claims of 100 for the January MOP. Assume that production starts in January and the reporting of claims is delayed according to the following lag probabilities:

$$\alpha_k = \begin{cases} 0.7, & k = 0 \\ 0.15, & k = 1 \\ 0.1, & k = 2 \\ 0.05, & k = 3 \\ 0, & \text{otherwise} \end{cases}$$

a. Give the expected number of claims for the months February, March, and April.

b. Determine the expected number of claims that will actually be reported in April if the average number of claims for each MOP is 100 and has these lag probabilities.

8. An automobile manufacturer released a new model of vehicle 16 months ago with a 3-year FRW policy. A MOP/MIS chart for the

warranty claims and production volumes, both in thousands of units, is given below.

		MIS													
MOP	VOL	1	2	3	4	5	6	7	8	9	10	11	12	13	14	15
1	10	5	4.8	4.7	5	4.7	3		3.2	2.9	2.9	2.3	2.2	2.4	2	2.1	1.9
2	10	4.9	4.8	4.65	6.2	5.9	4.2	3		2.8	2.9	2.6	2.4	2.3	2.1	1.9
3	10	4.2	4	4.1	3.8	3.9	3.6	3.3	3.29	2.87	2.49	2.38	2.4	2.3		
4	10	4	4.4	3.8	4.8	4.6	3.9	3.1	2.6	2.8	2.34	2.45	2.5			
5	10	3.6	4.3	3.81	3.9	3.1	3.2	2.9	2.8	2.65	2.69	2.63				
6	15	3.2	3.7	2.7	2.74	2.8	2.7	2.68	2.7	2.69	2.75					
7	15	2.9	3.8	2.9	2.85	2.8	2.82	2.78	2.83	2.8						
8	15	3.1	2.9	2.65	2.7	2.69	2.84	2.88	2.9							
9	15	3.1	2.9	2.8	3	2.76	2.9	2.8								
10	15	2.8	2.65	2.75	2.7	2.68	2.75									
11	10	2.7	2.5	2.8	2.4	2.55										
12	10	2.6	2.38	2.78	3.2											
13	10	2.6	2.41	2.8												
14	10	2.5	2.4													
15	10	2.3														

Construct a chart of the average number of defects per unit sold for each of 1-MIS, 6-MIS, and 12-MIS.

References

Groover, M.P., 2001, *Automation, Production Systems, and Computer-Integrated Manufacturing*, 2nd ed., Prentice Hall, Upper Saddle River, NJ, Chap. 4.

Kalbfleisch, J.D., J.F. Lawless, and J.A. Robinson, 1991, Methods for the analysis and prediction of warranty claims, *Technometrics*, 33, 3: 273-285.

Keeney, Ralph L. and H. Raiffa, 1993, *Decisions with Multiple Objectives*, Cambridge University Press, New York.

Marschak, J., 1950, Rational behavior, uncertain prospects, and measurable utility, *Econometrica*, 18, No. 3: 111-141.

Ross, S.M., 1983, *Stochastic Processes*, John Wiley & Sons, New York, Chap. 3.

Thomas, M.U., 1997, A methodology for product performance assessment, *Prod. Qual. Manage. Front.-VI*, C.G. Thor, J.A. Edosomwon, R. Poupart, and D.J. Sumanth, Eds., Engineering and Management Press, Norcross, GA, 649-659. Thomas, M.U., 1998, Manufacturing system control through warranty planning, *Flexible Automation and Intelligent Manufacturing*, H. Migliore, S. Randhawa, W.G. Sullivan, and M.M. Ahmad, Eds., Begell House Inc., New York, 747-757.

Voland, G., 1986, *Control Systems and Modeling Analysis*, Prentice Hall, Upper Saddle River, NJ.

6
The Quality Improvement Process

6.1 Introduction

This chapter describes an approach and methods for incorporating reliability in product system design decisions and developing alternatives for improving quality through reliability improvement methods. Quality improvement is an ongoing continuous process and, as we discussed in previous chapters, the reliability element of quality is most significant in product design. It is, therefore, essential that quality improvement efforts for products that function operationally include reliability as a major focus.

The total product life (TPL) of a manufactured product described in Chapter 3 (see Figure 3.1) consists of three stages: product development, production and manufacturing, and product usage. Reliability design is accomplished during product development, but includes results and information from production and customer usage stages. During the design, reliability specifications are formulated, including the overall reliability target, which is then allocated throughout the product components. Each such alternative design is then analyzed relative to cost and performance, and analyzed relative to the potential causes and impact of failures and safety. Preproduction analysis from quality control, acceptance, and burn-in testing provide additional data that are incorporated into the design alternatives.

Most manufacturing and service organizations are engaged in some form of continuous quality improvement program. Often, it is mandated by International Standards Organization (ISO) standards for particular producers and, for many manufacturers, continuous quality improvement is essential in order to remain competitive. It is safe to assume that some type of quality improvement program is a significant part of the business strategy for all consumer product manufacturers. Most programs introduce planned improvements periodically on an annual or model milestone basis where the changes are scheduled at specified times in the product production cycle. The quality improvement process is quite simple and consists of the following:

1. Identify where the quality problems exist in the production and manufacturing system.

2. Develop options for eliminating the problems and improving overall quality.
3. Evaluate the improvement options and determine a set of goals for the period.

It should be noted that there are two distinct scenarios for product design; one being the design of a completely new product and the other to make modifications for an existing model. So, in terms of reliability, one situation involves preproduction analyses and evaluation to predict the impact of design, while the other is to develop design changes for the purpose of improving quality. The major difference between these two scenarios is the amount and quality of the data that is available during the initial phase of the design process. Both conditions involve an iterative process of identifying problems, developing improvement options, and establishing improvement goals.

6.2 Identifying Quality Problems

Quality feedback from customer and warranty sources provides indications of problems, but further analysis is necessary to identify the actual causes of failures and malfunctions. For complex products and production systems, this can require extensive trouble-shooting and diagnostic studies. The most common techniques for examining failures are Failure Mode and Effect Analysis (FMEA); its extension, Failure Mode, Effect, and Criticality Analysis (FMECA); and Fault Tree Analysis (FTA). FMEA is a systematic procedure for decomposing a product or part into its most basic elements for determining the source of failure. It is the required process by any supplier to companies that subscribe to QS-9000, the quality standard that supplements ISO-9000. FMECA has the further purpose of estimating the risk associated with the failure modes. The procedure for performing FMECA is maintained by MIL-STD-1629A. FTA is a graphical technique for determining the root cause of undesirable events that is generally conducted in conjunction with FMECA.

6.2.1 FMECA Process

The purpose of FMECA is to find the problems and develop options for making improvement. It focuses on reliability improvement and is generally performed during the preliminary design phase of product development. The process consists of an iterative procedure of the following:

1. Identify all modes of failure
2. Conduct a failure risk analysis
3. Isolate the causes
4. Determine the corrective action for improvement

The Quality Improvement Process

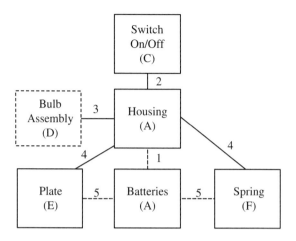

FIGURE 6.1
FMEA block diagram of a flashlight.
Reprinted from the Potential Failure Mode and Effects Analysis (FMEA) Manual with permission of DaimlerChrysler, Ford, and GM Supplier Quality Requirements Task Force.

The first step in the process is to identify the system including those elements or components likely to fail. Use various descriptive methods as appropriate, such as functional and process analysis techniques, block diagrams, and process flow charts. To illustrate the procedure, we consider the following example.

Example 6.1

Consider the block diagram of a flashlight taken from the Potential Failure Mode and Effects Analysis Reference Manual developed by Chrysler, Ford, and General Motors (ASQC/AIAG Task Force 1995) shown in Figure 6.1.

The flashlight is configured by five elements through the various connections listed in Table 6.1. To determine the failure modes, each linkage is examined relative to its source of failure. The numbers over the arcs in the diagram correspond to the type of attachment used for each of the elements. For improvement studies, additional information from quality feedback reports on warranties, customer complaints, and surveys can be useful in developing options for design changes.

TABLE 6.1

Flashlight Example for Potential Sources of Failure

Components:	Attaching Method:
A. Housing	1. Skip fit
B. Batteries (2 D Cell)	2. Rivets
C. On/off switch	3. Thread
D. Bulb assembly	4. Snap fit
E. Plate	5. Compressive fit
F. Spring	

Some common modes of failures for electromechanical devices are capacity overload, mechanical stress, contamination, fatigue, and friction. Some of the common causes of failures are due to the following:

- Defective parts
- Operator-induced error
- Corrosion
- Cracks
- Leaks
- Short circuits
- Fracture

The potential effects of failure are the effects of the failure mode on the respective function as it is perceived by the customer. So, it is important that effects are described in terms that a customer might experience or notice. For example, paint scars don't necessarily cause actual failure, but they lead to poor appearance that is not acceptable to customers. Also, any effects that impact safety or the threat of noncompliance with regulations are noted accordingly.

The risk analysis involves a complete review of the failure events and the relative impact on the system. Each failure mode is classified according to its risk through a classification of severity, assigned a relative probability of occurrence, and quantified in terms of a criticality index. The classification and impact procedure is shown in Table 6.2, ranging from Class IV failures that have little effect on performance to Class I failures that are catastrophic and life threatening.

Probability estimates are determined from the information that is available. Initially, estimates are based on known results from historical data, which is later amplified as necessary through reliability testing. In the absence of any information, the probabilities of occurrence are commonly taken from accepted guidance from Military Standards for estimating subjective probabilities provided by military standards such as MIL-HDBK 217F for electronic equipment. Criticality indices are then computed to establish an ordinal relationship among the failure modes, which provides a convenient means for classifying the risk and for planning corrective actions.

TABLE 6.2

Severity Classification for Failures

Class		Impact
I	Catastrophic	Major system failures with possible loss of life, injury, or major damage
II	Critical	Loss of system and performance is unacceptable
III	Marginal	System degraded and partial loss in performance
IV	Negligible	Minor failure with no affect on system performance

TABLE 6.3
Basic Fault Tree Symbols

Symbol	Name	Description
○	Fault Event	Independent primary fault tree event
▭	Fault Event	Result of logical combination of other events
AND gate symbol	AND Gate	Intersection operation of events whereby output event occurs if and only if all events occur
OR gate symbol	OR Gate	Union operation of events whereby output event occurs if one or more input events occur
◇	Fault Event	Event not fully developed due to causes being partially or completely unknown but assumed to be a primary fault event

6.2.2 Fault Tree Analysis

Fault Tree Analysis (FTA) is a graphical technique for determining the root cause of undesirable events or faults. This method focuses on faults and consists of identifying the combinations of events that will result in the occurrence of the most significant failure event — the top event. The accepted standard symbolic notation for representing fault tree events, adapted from Roberts et. al. (1981), is given in Table 6.3.

To illustrate the construction of a fault tree consider a simple alarm clock represented by the diagram in Figure 6.2. The clock consists of the four components: a power supply, timer, alarm, and switching mechanism for turning the alarm on and off. The power supply element has built-in standby

FIGURE 6.2
A simple electric alarm clock.

redundancy for current flow to the clock provided by a battery that is activated whenever the main source of electric power fails due to a power outage or surge in the line.

The fault tree for this clock is given in Figure 6.3. The top event is the failure of the alarm clock, which can occur due to any of the events, including A: loss of power input, B: clock element failure, or C: alarm failure. A loss of power to the clock can occur from event D: an electric power failure that can be the result of either of the events F: a power outage, G: a line surge in power, or H: a downed power line. However, there will still be a flow of current unless event E, the battery providing the back-up source of power also fails. An alarm failure will result from I: failure of the switch that activates the alarm, J: failure of the alarm mechanism, or K: the user sets the alarm incorrectly. This latter event might be considered a secondary failure event that is based on limited information or even with a high degree of uncertainty in the actual influence of human error in the cause of failure. Nonetheless, it can be a significant cause of quality problems and warranty expenditures if customers have difficulty operating the settings on the clock.

Once the failure tree is constructed, the next step is to determine the failure probability. Failure will occur, of course, when the top event occurs, and to determine this probability, we proceed from top event T in Figure 6.3, applying Boolean notation

$$T = A \cup B \cup C$$

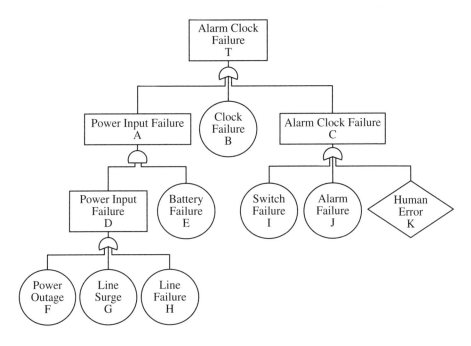

FIGURE 6.3
Fault tree of an electric alarm clock with backup battery power.

The Quality Improvement Process

Continuing downward and substituting for the resultant events as appropriate,

$$T = (D \cap E) \cup B \cup (I \cup J \cup K)$$
$$= [(F \cup G \cup H) \cap E] \cup B \cup (I \cup J \cup K)$$

It then follows that

$$P\{T\} = P\{(F \cap E) \cup (G \cap E) \cup (H \cap E) \cup B \cup I \cup J \cup K\}$$

For simple failure trees like this one, P{T} can be easily computed by direct expansion of the resultant events for the tree. For larger and more complex trees, this direct approach can be quite cumbersome. The common method for dealing with these failure trees is by minimum cut sets.

6.2.2.1 Procedure for Generating Cut Sets

A *cut set* is a collection of basic events that will cause the top event to occur and a *minimum cut set* is a cut set containing only the necessary events that will cause failure. The procedure consists of reducing the fault tree to a structure of the general form shown in Figure 6.4. The minimum cut sets can be generated through an iterative top down expansion procedure as follows:

1. Expand events input to OR gates by generating new rows.
2. Expand events input to AND gates by generating new columns.

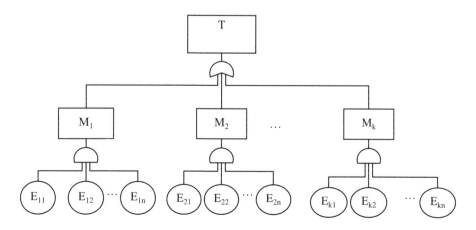

FIGURE 6.4
Minimum cut set fault tree structure.

3. Continue expansion until only basic or incomplete events remain.
4. Terminate when each row results in a cut set, and then eliminate the redundant cut sets.

Given the failure tree is in the structural form of Figure 6.4, then the top event

$$T = M_1 \cup M_2 \cup \ldots \cup M_k \tag{6.1}$$

is the union of the minimum cuts

$$M_i = E_1 \cap E_2 \cap \ldots \cap E_{n_i} \tag{6.2}$$

each containing the relevant basic events E_j. The probability of the top event, T, occurring is given by

$$P\{T\} = P\left(\bigcup_{i=1}^{k} M_i\right)$$

for which, in general

$$P\{T\} = \sum_{i=1}^{k} P\{M_i\} - \sum_{i=2}^{n}\sum_{r=1}^{i-1} P\{M_i \cap M_r\} \\ + \sum_{i=1}^{k}\sum_{r=2}^{i-1}\sum_{s=1}^{r-1} P\{M_i \cap M_r \cap M_s\} - \cdots + (-1)^{n-1} P\{M_1 \cap \ldots \cap M_n\} \tag{6.3}$$

and, for the minimum cut probabilities

$$P\{M\} = P\left(\bigcap_{j=1}^{n} E_{ij}\right) \tag{6.4}$$

The following three examples illustrate this reduction procedure for failure tree analysis.

Example 6.2

We will first consider the fault tree for the electric alarm clock given in Figure 6.3. The results for applying this method to this example are summarized in Table 6.4. Starting with the first level below the top event in Figure 6.3, we have the events A, B, and C. Since A and C are resultant events, we proceed to a second iteration. A can be expanded by its basic events consisting of the occurrence of both events D and E, and C can be expanded by occurrence of any of events I, J, or K.

TABLE 6.4
Minimum Cut Sets for the Electric Alarm Clock

Iteration	Tree cuts:				
1	A	B	C		
2	D,E	B	I,J,K		
3	F,E	G,E	H,E	B	I,J,K

Since D is a resultant event, a third iteration is taken by expanding it into its basic events F, G and H, any of which will lead to a power input failure if the back-up battery E also fails. Since there are no further resultant events, the final cut sets for the alarm clock are: $\langle (F,E),(G,E),(H,E),B,(I,J,K)\rangle$.

Example 6.3

As a second example, we will consider the fault tree for a vehicle brake system shown in Figure 6.5. The system consists of independent front and rear brake subsystems A and B, both of which must fail for the system to fail. Each subsystem failure can occur through failure of the wheel cylinders, brake linings, master cylinders, or insufficient brake fluid. A wheel cylinder failure can occur in the left or right wheel of the vehicle.

Applying the top-down procedure for determining the minimum cut sets, we start with a brake failure as the top event. The results are summarized in Table 6.5. Moving downward from the top brake failure event T, events A and H are resultant events, so we proceed to iteration 2. At iteration 2, the two events are expanded into (B, C,D,E) and (I, J K, L). At iteration 3,

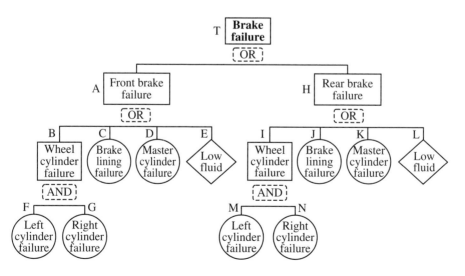

FIGURE 6.5
Minimum cut set fault tree structure for a fluid brake system.

124   Reliability and Warranties

TABLE 6.5

Minimum Cuts for the Brake System Example

Iteration	Tree Cuts:	
1	A	H
2	B,C,D,E	I,J,K,L
3	(F,G),C,D,E	(M,N),J,K,L

events B and I are further expanded into (F, G) and (M, N), thus resulting in the final cut sets:

$$\langle [(F,G),C,D,E] \cup [(M,N),J,K,L] \rangle$$

Assuming these events in Figure 6.4 are all independent, the failure probability for the top event, from (6.3) and (6.4), is

$$P\{T\} = P(F,G) + P(C) + P(D) + P(E)$$
$$+ P(M,N) + P(J) + P(K) + P(L)$$

or

$$P\{T\} = P(F)P(G) + P(C) + P(D) + P(E)$$
$$+ P(M)P(N) + P(J) + P(K) + P(L)$$

Example 6.4

As a final example, consider the fault tree for an alarm system from Ebeling (1997). This alarm could be a simple unit linked to a door or entryway for monitoring the security for a facility. The physical unit is a set of sensing devices located at strategic positions around the entryway and powered electrically with an auxiliary power supply that provides backup in case the primary power fails. The basic events for this system and their relationships are shown in Figure 6.6.

The results of the top-down procedure for generating minimum cut sets are summarized in Table 6.6. Starting at the first level below the top event with iteration 1, we have elements A, B, C, and D. Since A and B are resultant events, we proceed to the second iteration. We expand A by its two basic events, E and F, since failure will occur for that cut only if both primary and backup power sources fail. Similarly in B, both G and H must fail: however, G is a resultant event itself, so further reduction is necessary. Continuing the process for a third iteration leads to the final cut sets $\langle (E,F),(I,H),(J,H),(K,H),C,D \rangle$.

The failure probability for the alarm failure

$$P\{T\} = P(E)P(F) + P(H)[P(I) + P(J) + P(K)]$$
$$+ P(C) + P(D)$$

The Quality Improvement Process

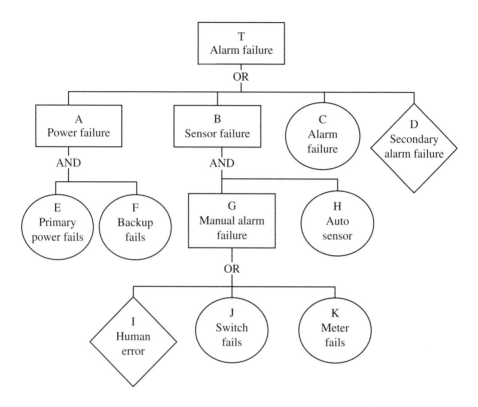

FIGURE 6.6
Example of a fault tree for an alarm system from Ebeling (1997).

for the case where all events are independent. If this is not the case and there are dependencies among some events, then it can be shown from Equation (6.3) and Equation (6.4) that this is a lower bound for P{T}.

6.3 Developing Quality Improvement Goals

Once the areas for improvement are known with the key problem areas identified, the next step is to find ways of improving reliability and overcoming the obstacles leading to failure. Reliability (being a major element of

TABLE 6.6

Minimum Cuts for the Alarm System

Iteration	Tree cuts:				
1	A	B	C	D	
2	E,F	G,H	C	D	
3	E,F	I,H	J,H	K,H	C D

quality) is routinely reviewed for product improvement and related reductions in warranty cost. This requires recurring examination of the parts and materials, new technologies for improving the control and reducing variation in the processes, and load conditions that can lead to failures. From these detailed analyses of each subsystem, various options can be identified that have to be examined for their feasibility relative to the business strategy. It, therefore, becomes necessary (for multiple component products) to establish guidance on determining goals for individual components that comply with an overall product system goal.

Reliability allocation is the process of translating the reliability requirements among the components of a system to achieve an overall system specification. This process can be difficult when designing new products where information on the operational and failure characteristics of the components are completely unknown. For the purpose of improving the quality of an existing production item, there is less uncertainty about component failure characteristics, but there are still challenging decisions in selecting and assigning component improvement targets. The basic problem is to find, for a fixed $t \geq 0$, functions $g_i[R_1(t),...,R_n(t)]$ that assigns an apportioned reliability improvement for each component i having current reliability $R_i(t)$ to achieve an overall reliability of R(t). The early methods for reliability allocation emerged from the Department of Defense's electronics industry ARINC (1964). The product system is treated as a series configuration of n independent components, each functioning during its useful life. The system reliability is then

$$R(t) = e^{-t\sum_{i=1}^{n}\lambda_i}, \quad t \geq 0 \qquad (6.5)$$

where λ_i is the failure rate that is constant for component i = 1,...,n. The simple approach for allocating improvement goals among the n components is to find a function g(.) that assigns target weights to each component based on criteria, such as relative importance, difficulty in achieving improvement in the component, or cost for implementing the changes to accomplish the improvement. A popular method adopted by the Advisory Group on Reliability of Electronics Equipment (AGREE) in 1957 established subsystem or component goals by incorporating complexity factors based on the number of parts and the importance factors based on the conditional failure probability given a subsystem failure.

The AGREE method and its variants do not consider the costs associated with failures or for implementing the changes for improvement. Cost considerations are essential in planning improvements in product manufacturing. Below is the description of a method by Thomas and Richard (2006) for allocating reliability improvement efforts where the impact on quality and cost effectiveness will be greatest relative to customer and market preferences.

The Quality Improvement Process

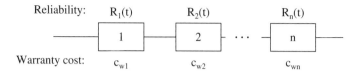

FIGURE 6.7
An n-component product with reliability and warranty cost $R_i(t)$ and c_{wi}.

6.3.1 Thomas–Richard Method

Consider a product consisting of a series arrangement of n components with reliabilities $R_i(t)$, $i = 1,\ldots,n$, shown in Figure 6.7 that is sold under a warranty of length T_w. For convenience, one can think of an FRW policy, though it doesn't matter as long as the length and the manufacturer's cost, c_{wi}, for warranty claims made during the warranty period are known. We further assume the following:

1. Failure times are independent and result in warranty claims.
2. The planning horizon for evaluating reliability improvements is the length of the warranty period, T_w.
3. The improved product will have independent components that have exponentially distributed failure times.

Our objective is to determine the n component reliabilities, $R_i^0(T_w)$, for accomplishing a 100γ, $0 < \gamma < 1$ percent aggregate improvement over the current component levels, $R_i(T_w)$, $i = 1,\ldots,n$. The product system goal is, therefore,

$$R^0(T_w) \geq (1+\gamma)R(T_w) \tag{6.6}$$

The expected cost of component i failures during a warranty period $[0, T_w]$ is $c_{wi}M_i(T_w)$ and for the product system

$$C(T_w) = \sum_{j=1}^{n} c_{wj}M_j(T_w) \tag{6.7}$$

where $M_j(T_w)$, from Equation (2.44) in Chapter 2, is the expected number of claims occurring in $[0, T_w]$ that was discussed in Section 2.3. We can define the warranty burden for component i

$$\beta_i = \frac{c_{wi}M_i(T_w)}{\sum_{j=1}^{n} c_{wj}M_j(T_w)} \tag{6.8}$$

which is the fraction of the total warranty cost that is contributed by component i. Note that the warranty cost for a component provides a relative indicator of its quality and components that fail frequently or have high warranty expense or both will have correspondingly high warranty burden. Therefore, under an improvement initiative the goal should be to reduce failures at each component i in proportion to the fraction of warranty cost due to that component. Clearly, those components with higher values of β_i should receive larger reduction targets than those with less warranty burden. Thus,

$$\frac{M_i(T_w) - M_i^0(T_w)}{M(T_w) - M^0(T_w)} \geq \beta_i, \quad i = 1, \ldots, n \tag{6.9}$$

and it follows that

$$M_i^0(T_W) = \begin{cases} M_i(T_W) - \beta_i[M(T_W) - M^0(T_W)], & \frac{M_i(T_W)}{\beta_i} \geq M(T_W) - M^0(T_W) \\ M_i(T_W), & \frac{M_i(T_W)}{\beta_i} < M(T_W) - M^0(T_W) \end{cases} \tag{6.10}$$

By assumption 3, once the changes are implemented, the failure times for the improved components will be distributed exponentially. It, therefore, follows from the discussion in Chapter 2, Section 2.3 that the number of failures during the warranty period will be $\lambda_i^0 T_w$ for component i and $\lambda^0 T_w$ for the product. In Equation (6.10), it then follows that the average number of failures for the components is given by

$$\lambda_i^0 T_W = \begin{cases} M_i(T_W) - \beta_i\left[M(T_W) - \lambda^0 T_W\right], & \frac{M_i(T_W)}{\beta_i} \geq M(T_W) - \lambda^0 T_W \\ M_i(T_W), & \frac{M_i(T_W)}{\beta_i} < M(T_W) - \lambda^0 T_W \end{cases} \tag{6.11}$$

We will consider two cases for reliability allocation: (1) when the product currently has a constant failure rate (CFR) and (2) the situation where the failure rate is increasing (IFR).

6.3.1.1 Constant Failure Rate (CFR) Allocation

Given a product that currently functions in its useful life during $[0, T_w]$ with component failure rates $\lambda_1, \ldots, \lambda_n$, the objective is to determine new rates $\lambda_1^0, \ldots, \lambda_n^0$ such that the improved reliability will satisfy Equation (6.6). Since the reliability of the improved product is to be

$$R^0(T_w) = e^{-T_w \sum_{i=1}^{n} \lambda_i} \tag{6.12}$$

The Quality Improvement Process

it follows from Equation (6.6) that

$$\sum_{i=1}^{n} \lambda_i^0 \leq \frac{1}{T_w} \ln\left[\frac{1}{(1+\gamma)R(T_w)}\right]$$

Since the number of claims in $[0, T_w]$ is distributed Poisson with the expected number of claims $M_i(T_w) = \lambda_i T_w$, for $i = 1,\ldots,n$, then from Equation (6.9)

$$\frac{\lambda_i - \lambda_i^0}{\lambda - \lambda^0} \geq \beta_i, \, i = 1,\ldots,n,$$

where

$$\lambda = \sum_{i=1}^{n} \lambda_i$$

is the aggregate failure rate for the current product and, similarly, λ^0 is the aggregate rate for the improvement goal, and further Equation (6.11) reduces to

$$\lambda_i^0 = \begin{cases} \lambda_i - \beta_i(\lambda - \lambda^0), & \lambda_i/\beta_i \geq \lambda - \lambda^0 \\ \lambda_i, & \lambda_i/\beta_i < \lambda - \lambda^0 \end{cases},$$

since the T_w terms cancel out. We will illustrate this procedure through the following example.

Example 6.5

A gearbox assembly for a motor vehicle described by Ivanovic (2000) consists of a series configuration of the following five subsystems:

1. Box with components
2. Input and main shaft with internal gears, synchronizer assembly, and bearing
3. Counter and reverse idler shaft with internal gears, synchronizer assembly, and bearing
4. Selector forks with components
5. Gear selector mechanism

We assume that these subsystems are independent and with constant failure rates and marginal warranty costs given in Table 6.7.

The gearboxes are sold with a 1-year free replacement warranty. The company would like to improve the overall reliability of this product by 20%.

TABLE 6.7

Gearbox Assembly with Subsystem Failure Rates and Warranty Costs

Subsystem	Description	Failure Rate λ_i	Warranty Cost c_{wi} ($/unit)
1	Box with components	0.02737	7.37
2	Input and main shaft with internal gears, synchronizer assembly, and bearing	0.07796	10.42
3	Counter and reverse idler shaft with internal gears, synchronizer assembly, and bearing	0.11093	9.58
4	Selector forks with components	0.07668	5.16
5	Gear selector mechanism	0.24207	8.23

Summing the subsystem failure rates in Table 6.6, the failure rate for the gearbox is $\lambda = 0.53501$. The subsystem failure rates are assumed to remain constant over the $T_w = 1$-year warranty period, therefore the reliability is

$$R(1) = e^{-0.53501(1)} = 0.58566$$

So, for a 20% improvement, $\gamma = 0.2$ and the target from Equation (6.6) is to achieve

$$R^*(1) \geq (1.2)(0.58566) = 0.70279$$

or from Equation (6.12) reduce the failure rate to

$$\lambda^* \leq \ln\left(\frac{1}{0.70279}\right) = 0.35270$$

The total expected warranty cost is

$$\sum_{i=1}^{5} c_{wi} \lambda_i T_w = 7.37(0.02737) + \cdots + 8.23(0.24207) = 4.4647$$

since $T_w = 1$. The warranty burden for subsystem-1 is then computed from Equation (6.8)

$$\beta_1 = \frac{(7.37)(0.02737)}{4.4647} = 0.04518$$

The Quality Improvement Process

and, similarly, $\beta_2 = 0.18195$, $\beta_3 = 0.23803$, $\beta_4 = 0.08862$, and $\beta_5 = 0.44622$. Substituting these rates into Equation (6.14), the target failure rate for subsystem-1 is

$$\lambda_1^0 \leq 0.02737 - (0.04518)(0.53501 - 0.35269) = 0.01913$$

since $\lambda_1/\beta_1 > 0.18232$. The remaining target values follow with $\lambda_2^0 = 0.04479$, $\lambda_3^0 = 0.06753$, $\lambda_4^0 = 0.06052$, and $\lambda_5^0 = 0.16072$.

6.3.1.2 Increasing Failure Rate (IFR) Allocation

Sometimes even during the warranty period one or more of the components of a product can have an increasing failure rate (IFR). This might occur when an item is placed on the market before it has been properly tested or there is a problem with a vendor that supplies critical components. So, for this situation in Equation (6.12), the objective is to develop improvement goals that will provide a reliability of

$$R^0(T_W) = (1+\gamma)e^{-\int_0^{T_W} \lambda(u)du} \tag{6.13}$$

To determine the improvement goals from Equation (6.11), we have to determine the values for $M_i(T_w)$ for $i = 1,\ldots,n$.

Consider a three-component system with two that are CFR and one has a linearly increasing hazard rate. Let $\lambda_1(t) = \lambda_1$, $\lambda_2(t) = \lambda_2 t$, and $\lambda_3(t) = \lambda_3$, for $t \geq 0$; thus, we have for components 1 and 3, $M_1(t) = \lambda_1 t$ and $M_3(t) = \lambda_3 t$. To establish $M_2(t)$, we determine the probability density function for the time between occurrences of warranty claims from

$$f_2(t) = -\frac{d}{dt} R(t) = \lambda_2 t e^{-\lambda_2 t^2/2}, \quad t \geq 0 \tag{6.14}$$

which is the Erlang probability density function. The Erlang is a comman distribution in reliability and queuing theory, arising from sums of independent exponentially distributed random variables. It follows that the renewal function (see Appendix C) is

$$M_2(t) = \frac{1}{4}e^{-2\lambda_2 t} + \frac{\lambda_2}{2}t - \frac{1}{4}, \quad t \geq 0 \tag{6.15}$$

The renewal function for the product is, therefore,

$$M(T_W) = \frac{1}{4}e^{-2\lambda_2 T_W} + \left(\lambda_1 + \frac{\lambda_2}{2} + \lambda_3\right)T_W - \frac{1}{4}, \quad t \geq 0 \tag{6.16}$$

Substituting this equation for $M(T_w)$ into Equation (6.8), we get the warranty burden rates

$$\beta_1 = \frac{c_{W1}\lambda_1 T_W}{c_{W1}\lambda_1 T_W + c_{W2}\left(\frac{1}{4}e^{-2\lambda_2 T_W} + \frac{\lambda_2}{2}T_W - \frac{1}{4}\right) + c_{W3}\lambda_3 T_W} \tag{6.17}$$

$$\beta_2 = \frac{c_{W2}\left(\frac{1}{4}e^{-2\lambda_2 T_W} + \frac{\lambda_2}{2}T_W - \frac{1}{4}\right)}{c_{W1}\lambda_1 T_W + c_{W2}\left(\frac{1}{4}e^{-2\lambda_2 T_W} + \frac{\lambda_2}{2}T_W - \frac{1}{4}\right) + c_{W3}\lambda_3 T_W} \tag{6.18}$$

and

$$\beta_3 = \frac{c_{W3}\lambda_3 T_W}{c_{W1}\lambda_1 T_W + c_{W2}\left(\frac{1}{4}e^{-2\lambda_2 T_W} + \frac{\lambda_2}{2}T_W - \frac{1}{4}\right) + c_{W3}\lambda_3 T_W} \tag{6.19}$$

These rates are then substituted into Equation (6.11) to determine the failure rate improvement goals.

Example 6.6

Suppose for the three-component product described above, $\lambda_1 = 0.1$, $\lambda_2 = 0.15$, and $\lambda_3 = 0.05$ with a warranty of length $T_w = 1$ year. The component warranty costs are $c_{w1} = c$, $c_{w2} = 1.5c$, and $c_{w3} = 0.5c$ units of cost in terms of a constant $c > 0$. The overall goal is to improve the reliability by 20%.

The current reliability for this product is given by

$$R(T_W) = e^{-[\lambda_1 T_W + (\lambda_2/2)T_W^2 + \lambda_3 T_W]}$$

and, on substituting the parameter values, we have

$$R(1) = e^{-1[.1(1)+(.15/2)(1)^2+.05(1)]} = 0.7985$$

and, from Equation (6.13), the overall product goal is $R^0(1) = (1.2)(0.7985) = 0.9582$. The mean number of claims for each of the components is $M_1(1) = (0.1)(1) = 0.1$, $M_3(1) = (0.05)(1) = 0.05$, and from Equation (6.15)

$$M_2(1) = \frac{1}{4}e^{-2(.15)1} + \frac{.15}{2}(1) - \frac{1}{4} = 0.0102$$

and the total mean number of claims during the 1-year period is $M(1) = 0.1602$. Substituting these values into Equation (6.17), we have

$$\beta_1 = \frac{.1(1)}{.1(1) + 1.5\left[\frac{1}{4}e^{-2(.15)1} + (.15/2)(1) - 1/4\right] + .5(.05)1} = 0.7128$$

and from Equation (6.18) and Equation (6.19), $\beta_2 = 0.1091$ and $\beta_3 = 0.1782$. So, for an improved product reliability of $R^0(1) = 0.9582$, it follows from Equation (6.12) that $\lambda^0 = 0.0427$. Therefore, applying Equation (6.11), we have

$$M_1^0(1) = 0.1 - .7127[.1602 - .0427] = 0.0162 \; ; \; M_2^0(1) = 0.0102$$

since $M_2(1)/\beta_2 < M(1) - M^0(1)$; and $M_3^0(1) = 0.0367$.

The solution is, therefore, $\lambda_1^0 = 0.0162$, $\lambda_2^0 = 0.0102$, and $\lambda_3^0 = 0.0367$.

Note that the solution is to implement changes to improve components 1 and 3 without pursuing improvement for component 2. This is because even though component 2 has an increasing failure rate and a large warranty burden, the expected number of failures during the 1-year warranty period is small relative to the other two components. However, if the warranty period is extended 3 months to $T_w = 1.25$, then all three components would be targeted for improvement.

This method provides preliminary goals that serve as targets requiring further refinement at the subsystem or component level to explore their technical and economical feasibility. The preliminary analysis might provide a logical goal for focus on a particular component that understandably would improve the reliability and, hence, the quality of the product, but the cost to make the necessary modification is too high for the financial conditions of the organization. A number of optimization procedures have been developed for dealing with reliability optimization problems in general. Details on the mathematical programming methods for solving these models under a range of structural conditions on the system configuration and formulation can be found in Kuo and Zuo (2003).

6.4 Decision Analysis Framework

The process of analyzing and evaluating design improvement options generally involves cost trade-off decisions relative to quality and reliability. The amount of information available in making these decisions can vary widely from components that have received considerable testing and, therefore, have known failure characteristics to the introduction of a new product for which only limited information is available for making evaluations. Formally, decision analysis is a structured process for evaluating alternatives. Essentially all decision problems can be formulated through a standard decision analysis framework consisting of the triplet specifying {**A**, **Θ**, **V**} where

$\Theta \equiv$ set of future outcome states of nature: $(\theta_1, \theta_2, \ldots, \theta_n)$

$A \equiv$ set of alternatives: $(a_1, a_2, \ldots, a_m)$

$\mathbf{V} \equiv$ consequence values: $(v_{ij} = V(a_i, \theta_j); \; i = 1, \ldots, m; \; j = 1, \ldots, n)$

TABLE 6.8

Decision Analysis Matrix for m Alternatives and n States of Nature.

Alternatives	Outcomes of Nature:			
	θ_1	θ_2	...	θ_n
a_1	$v_{1,1}$	$v_{1,2}$	...	$v_{1,n}$
a_2	$v_{2,1}$	$v_{2,2}$	...	$v_{2,n}$
...	...	...	...	...
a_m	$v_{m,1}$	$v_{m,2}$	...	$v_{m,n}$

For the case of the states of nature Θ being discrete, the decision problem can conveniently be displayed in the matrix format of Table 6.8. The common metrics for the consequences are cost, profit, number of failures, or cost per failure. We start with the following assumptions:

1. The set of consequence values (v_{ij}: i = 1, ...,m; j = 1,...,n) are known.
2. Θ can be specified but the outcomes of θ_j can be random.
3. Combinations of (a_i, θ_j) are independent, i.e., outcomes of Θ are not influenced by choices of a_i.

In order to determine the consequence values v_{ij} and select a choice from among the m alternatives, it is necessary to establish the criteria by which the decision is to be based. The most common consequence measures are cost or profits, but others, such as reliability, warranty costs, or marketing costs, can be used as well. It is, of course, necessary that all v_{ij} values are assessed consistently in determining the alternative that provides the best solution. It is the level of knowledge about Θ that determines whether a decision problem is one under assumed certainty, risk, or complete uncertainty.

6.4.1 Decisions under Risk Conditions

The decision analysis approach is useful in analyzing reliability design decisions because it provides a structure for separating facts from subjective inputs, and applying a rationale. For most reliability design decisions, there is some knowledge about the failure characteristics, which generally includes the failure time distribution. When the probabilities of occurrence of the states of nature are known, the decision problem is one involving risk and the alternatives can be evaluated by an expectation principle. The most common criterion in reliability analysis is to select the alternative that will result in the minimum expected costs or maximum expected profit, subject to a set of derived constraints on the reliability. This method is illustrated in the following example.

Example 6.7

Let us revisit the gearbox assembly of Example 6.2 with the subsystems comprised of the subsystems:

1. Box with components
2. Input and main shaft with internal gears, synchronizer assembly, and bearing
3. Counter and reverse idler shaft with internal gears, synchronizer assembly, and bearing
4. Selector forks with components
5. Gear selector mechanism

Suppose that after further analysis of the financial options for improving the gearbox assembly the following alternative actions have been identified.

a_1: Upgrade subsystems 3, 4, and 5
a_2: Upgrade subsystems 1, 2, 3, and 4
a_3: Do not upgrade the assembly this year

Alternative a_1 was recommended based on the preliminary analysis. However, the proposed changes for the gearbox selector mechanism require a new sensor element that could cause excessive warranty expense if it fails to meet product life expectations. Alternative a_2 requires the same investment of $280,000 as a_1 and will affect four of the five subsystems. The impact on these alternatives will depend on the market, which is driven by the prevailing economic conditions during the forthcoming year. This impact also applies to alternative a_1, which is to do nothing in reducing failure rates this year. After considerable review of past data and forecasts, the company has identified the following potential outcomes that can result from the economy for the next year.

θ_1: Pessimistic view; sales reduction by 50%
θ_2: No change from the current year
θ_3: Optimistic view; sales will increase by 50%

The resulting consequences, v_{ij}, of these states of nature in the form of estimated equivalent annual cost are given in Table 6.9. The equivalent annual

TABLE 6.9
Decision Consequences for Example 6.7

Alternatives	Nature Outcome Probabilities		
	$P(\theta_1) = 1/4$	$P(\theta_2) = 1/2$	$P(\theta_3) = 1/4$
a_1	80	100	150
a_2	75	120	100
a_3	60	110	180

cos,t which includes the cost for warranty expense, maintenance and repair, and product modifications is determined for each alternative and outcome 1, 2, and 3.

For this example the distribution for the states of nature, i.e., the future sales conditions, are assumed to be $P(\theta) = (1/4, 1/2, 1/4)$.

Applying the minimum expected equivalent annual cost criteria, we seek the alternative

$$\min_i EAC(a_i) = \min_i \sum_{j=1}^{3} V(a_i, \theta_j) P(\theta_j), \quad i = 1, 2, 3 \quad (6.20)$$

where $P(\theta_j)$ are the known probabilities of outcomes of Θ. Here the equivalent annual costs are

$$EAC(a_1) = \frac{1}{4}(80) + \frac{1}{2}(100) + \frac{1}{4}(150) = 107.50 \text{ \$ units/year}$$

$$EAC(a_2) = \frac{1}{4}(75) + \frac{1}{2}(120) + \frac{1}{4}(100) = 103.75$$

$$EAC(a_3) = \frac{1}{4}(60) + \frac{1}{2}(110) + \frac{1}{4}(180) = 115.00$$

for which the minimum occurs with alternative a_2. So, based on these conditions and criterion, the rationale decision would be to upgrade subsystems 1, 2, 3, and 4.

Example 6.8

A manufacturer produces a heavy terrain vehicle that is used in commercial construction and military operations. The primary source of failure for this vehicle is the servo controller that maintains the control and stability of the vehicle through feedback signals from a collection of sensors mounted on the drive train. Currently, the vehicles are sold with a 3-year FRW policy and the end of warranty reliability is 0.92. The company is being pressured to improve this reliability, and the warranty costs are becoming excessive.

The current method for connecting the sensors is by conventional metal fasteners that are welded to the frame of the vehicle, which detach over time due to vibrations. Two alternative designs have been developed for mounting the sensors to improve the reliability, but involve serious risk considerations. The following alternatives are available for consideration:

a_1 : Keep the current system with the sensors held by metal fasteners.

a_2 : Sensing information is transmitted through smart materials linked to the controller.

a_3 : Replace the current sensors with newly developed miniature sensors that are mounted by an anaerobic adhesive process.

The cost for implementing the changes are about the same for alternatives a_2 and a_3, but for a_2 the process will be more cost effective for military operations, whereas a_3 is more suitable for construction operations. The main difference is in warranty repairs. While improving the reliability of the vehicle is important, the company also faces risk in its future sales. Based on a market analysis, there is a 10% chance that the company will see an increase in sales from the construction industry due to a major competitor going out of business. However, there is also a 20% chance the government will impose major cutbacks in defense spending that could cause a loss in sales by as much as 50%. The states of nature for the decision problem, therefore, are as follows:

θ_1 : A 50% loss in government sales
θ_2 : No change in market conditions
θ_3 : A 50% loss in government sales and a 10% increase in construction sales
θ_4 : A 10% increase in construction sales

An analysis of the sales forecast and the performance data derived through testing and the different design options revealed the projected profits for the company under each alternative a_i and future outcome of nature θ_j; the values are given in Table 6.10 along with the estimated probabilities $P(\theta_j)$.

Applying the maximum equivalent annual profit criterion, we seek the alternative

$$\max_i \text{EAP}(a_i) = \max_i \sum_{j=1}^{4} V(a_i, \theta_j) P(\theta_j), \, i = 1, 2, 3 \qquad (6.21)$$

Therefore,

$$EAP(a_1) = 60(0.20) + 75(0.05) + 100(0.65) + 125(0.10) = \$93.75M$$

And similarly, $EAP(a_2) = \$96.75M$ and $EAP(a_3) = \$100M$. The best choice between the three alternatives based on the expectation principle is, therefore, a_3, invest in the miniature sensor improved design.

TABLE 6.10

Decision Consequences in $Million Sales for Heavy Terrain Vehicles

	Nature Outcome Probabilities:			
	$P(\theta_1) = 0.20$	$P(\theta_2) = 0.05$	$P(\theta_1) = 0.65$	$P(\theta_1) = 0.10$
a_1	60	75	100	125
a_2	50	45	110	130
a_3	50	35	115	135

6.4.2 A Generalized Maximum Entropy Principle (GMEP)

When the distribution about Θ is not known, then in order to apply an expectation criterion for evaluating the alternatives in **A**, it becomes necessary to apply a principle to establish the probabilities $p_1,...,p_n$. One option that has received wide acceptance as a rational approach for dealing with uncertainty is to derive a maximum entropy distribution based on the given facts about Θ. The maximum entropy concept emerged from early work in communications theory and was adopted as a principle in deriving distributions by Jaynes (1957). For discrete events $x_1, x_2,..., x_n$, defined on some sample space **S**, with probabilities $p_i = P(X = x_i)$, the entropy

$$H(p) = -\sum_{i=1}^{n} p_i \ln p_i \tag{6.22}$$

represents the amount of information necessary to know the outcome of X with certainty. For the trivial case of $p_i = 1$, H(p) = 0 denoting the certainty of outcome x_i. Often one might be able to make reasonable assumptions about properties of particular events and moments of random variables used in modeling, yet not be able to specify the actual distribution. Under these circumstances a maximization problem can be formulated to maximize H(p) in Equation (6.22) subject to a set of constraint conditions developed from the known factual data or information. This maximum entropy distribution can then be used in conjunction with the expectation principle described in the previous section, treating the problem as one involving risk. Thomas (1979) formulated this generalized maximum entropy principle as a nonlinear programming problem and applied the method to selecting oil spill recovery systems for cleaning inland harbors.

Let X represent a discrete random variable that takes on values $x_1,..., x_n$ with respective probabilities, $p_1,..., p_n$. The problem of selecting a distribution P() over is to solve the nonlinear program

$$\max_{p} H(p) = -\sum p_i \ln p_i$$

$$st: \sum_i p_i = 1, \ p_i \geq 0 \tag{6.23}$$

$$\alpha_{jk} \leq \sum_{i \in I_k} \xi_k(x_i) p_i \leq \beta_{jk}, \quad j = 1, 2,...; k = 0, 1,...$$

where I_k is the set of integers with α_{jk} and β_{jk} constants and $\xi_0(x_i) = 1$ for each i. Examples of functions $\xi_k(x_i)$ are $\xi_1(x_i) = x_i$, $\xi_2(x_i) = x_i^2$, and $\xi_3(x_i) = (x_i - \bar{x})^2$.

The solution to Equation (6.23) can be obtained by classical optimization methods. Let $g(p) = (g_1(p),..., g_r(p))$ be a vector representation of r constraints for an n-state discrete probability vector p. It follows that the necessary and sufficient conditions for p and a parameter vector γ to be a solution are

$$\nabla H(p) - \gamma g(p) = 0; \; \gamma_i g_i(p) = 0, \; i = 1,\ldots, r; \; \gamma \geq 0; \; \text{and} \; g(p) \leq 0 \quad (6.24)$$

Example 6.9
We will now apply the GMEP to our gearbox assembly problem using the data given in Table 6.9, less the values p_1, p_2 and p_3. Without any further information about Θ, from Equation (6.22) the problem is to

$$\max_p H(p) = -\sum_{i=1}^{3} p_i \ln p_i$$

$$\text{st}: \sum_{i=1}^{3} p_i = 1; \quad p_i \geq 0, \; i = 1, 2, 3$$

(6.25)

This problem can be solved by the calculus of variations from which the solution is $p_1 = p_2 = p_3 = 1/3$. The interpretation is that, given no more information beyond having a distribution Pθ, the outcomes θ_1, θ_2 and θ_3 are equally likely to occur. Applying this distribution to compute the equivalent annual costs for the alternatives

$$EAC(a_1) = \frac{1}{3}(80) + \frac{1}{3}(100) + \frac{1}{3}(150) = 110.00 \; \$ \; \text{units/year}$$

$$EAC(a_2) = \frac{1}{3}(75) + \frac{1}{3}(120) + \frac{1}{3}(100) = 98.33$$

$$EAC(a_3) = \frac{1}{3}(60) + \frac{1}{3}(110) + \frac{1}{3}(180) = 116.67$$

and a_2 is the minimum cost alternative.

Now suppose the company realizes that last year was a typical of economic conditions and it is highly unlikely that next year will be as bad. The leading indicators for the market suggest there is a 50% chance that their sales will double. Modifying Equation (6.24) to include this information, our optimization problem becomes

$$\max_p H(p) = -\sum_{i=1}^{3} p_i \ln p_i$$

$$\text{st}: \sum_{i=1}^{3} p_i = 1 \quad (6.26)$$

$$p_1 \geq 0, \; p_2 \geq 0, \; p_3 = 0.5$$

The solution to this program is $p_1 = 1/4$, $p_2 = 1/4$, $p_3 = 1/2$. Applying this distribution, we find that again the optimum choice is to select a_2, which has an expected equivalent annual cost of $98.75 units per car.

The maximum entropy principle has wide application in modeling and engineering decision-making problems involving risk for both discrete and continuous random variables. For more details on applying this method, see Tribus (1969).

6.4.3 Decisions under Complete Uncertainty

There are many situations where it is not practical to come up with a distribution over Θ. In other words, there is a lack of sufficient evidence or belief that any particular state of nature will occur. This is particularly true when the states of nature involve the economy. To analyze these problems, the decision maker needs to apply some criterion that reflects his or her attitude toward the prevailing risks and consequences. A common approach for dealing with these types of decision problems is to apply the *minimax principle* of choice for the case of the $V(a_i, \theta_j)$ representing costs, of *maximin principle* for the profit version of the decision problem. For cost, the criterion is to choose the alternative a_{i^*}, such that the cost

$$C(a_{i^*}, \theta_j) = \min_i \max_j C(a_i, \theta_j), \quad i = 1,\ldots,m;\ j = 1,\ldots,n$$

This principle is illustrated in the following example.

Example 6.10
Suppose for the gearbox assembly problem of Example 6.8 that the distribution of the future state of the sales conditions are completely unknown. Therein, the matrix of Table 6.9 remains as follows:

Alternatives	States of Nature:		
	θ_1	θ_2	θ_3
a_1	80	100	150
a_2	75	120	100
a_3	60	110	180

However, for these circumstances the probabilities over Θ cannot be specified. Applying the minimax criterion, we note that the row maximum for a_1 is 150, for a_2 it is 120, and for a_3 it is 180. The solution, therefore, is the alternative a_{i^*} for which

$$C(a_{i^*}, \theta_j) = \min_i (150, 120, 180) = 120,$$

The Quality Improvement Process

which implies that $a_{i^*} = a_2$ is the choice of alternative.

Note that for all of these scenarios the solution led to the same alternative. For this problem, the company would probably feel comfortable going with this option for improvements. The minimax principle is a common method that is applied for dealing with decision problems under uncertain conditions, though it is considered by some people to be too conservative. There are other methods as well that can be applied to dealing with decisions under uncertainty (Raiffa, 1970).

6.5 Exercises

1. Draw a reliability block diagram for the fault tree of the electric alarm clock in Example 6.3.
2. Draw a reliability block diagram for the fault tree of the alarm system in Example 6.6.
3. Find a fault tree for the system failure for each of the following configurations.

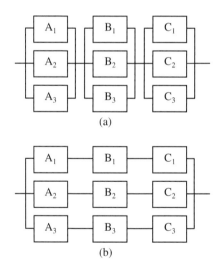

4. The schematic diagram in Figure 6.8 represents a laboratory faucet system consisting of a spout and hot and cold water assemblies that are connected to the water lines by connections at A and D. The remaining connector junctions B, C, E, and F provide the water flow to the spout assembly for each of the H and J water valves. All events represent sources of water leaks.
   a. Construct the fault tree for the system.

b. Find the minimum cut sets.

c. Assuming all connection failure events are independent, compute the probability of a system failure given the connector probabilities (Figure 6.8): $p_A = p_D = 0.025$, $p_B = p_E = 0.05$, $p_C = p_F = 0.10$, and $p_G = p_H = p_I = p_J = 0.01$.

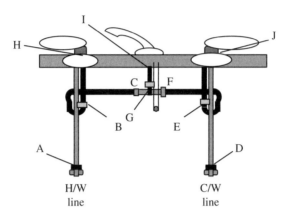

FIGURE 6.8
A laboratory faucet system.

5. Find minimum cut sets for the following fault tree.

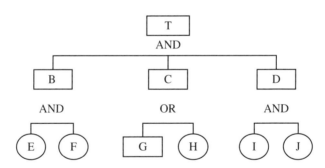

6. Determine the top event occurring for the fault tree in Exercise 5 given the following event probabilities: $p_E = p_F = 0.05$, $p_G = 0.15$, $p_H = 0.10$, and $p_I = p_J = 0.05$.

7. Determine the allocation for reliability improvement goals in Example 6.5 for an improvement objective of $\alpha = 0.10$.

8. The laboratory faucet system in Figure 6.8 consists of the following four components:

- Spout
- Hot water valve assembly with handle

- Cold water valve assembly with handle
- Connecting tube with T-junction connector

The failure rates and marginal warranty costs estimated from historical data are given as follows:

Subsystem	Failure Rate (failures/hr)	Warranty Cost ($/unit)
1. Water spout	0.005	15.00
2. Hot water valve assembly	0.05	10.00
3. Cold water valve assembly	0.05	10.00
4. Connecting tube assembly	0.10	7.50

Determine a set of reliability goals for each component given the manufacturer wants to improve the reliability of the system by 10%. Assume a warranty period of 1 year and the failures for each component are independent and distributed exponentially.

9. Suppose in Exercise 8 the failure times for the components 1, 2, and 3 are exponentially distributed, but for component 4, the failure time probability density function is given by,

$$g_4(t) = \lambda_4 t e^{-\lambda_4 t/2}, t \geq 0$$

Determine the reliability allocation among the components.

10. A company has identified three optional designs for making improvements in a product. The return on the investment for improvement for each alternative can have varying effects on the warranty cost depending on the forthcoming market conditions and risk in getting the improvements that are predicted. The estimates for the consequences of these options for the given future outcomes of poor, moderate, and good sales are given in the following decision matrix.

	Future Sales Conditions in $×1,000,000;		
	Θ: Poor	Moderate	Good
a_1	5.6	9.0	12.5
a_2	6.8	7.9	11.0
a_3	5.0	7.1	14.1

a. Determine the best choice from the alternatives using the expectation principle with known probabilities: $P(\theta_1) = 0.2$, $P(\theta_2) = 0.6$, and $P(\theta_3) = 0.2$.

b. Suppose you are determined to ensure that the choice you take will maximize the probability of sales of at least $7.5M.

c. Suppose the probabilities are not known, find the maximin solution.

11. Suppose the manufacturer of the faucet system in Exercise 8 is considering alternatives to the current method for connecting the water from the valve connections to the faucets. The current method uses copper tubing that is connected at B, C and E, F in Figure 6.8. The reliability is not good for this arrangement and building contractors complain about the material cost and the installation time for connecting the tubes. Alternatively, one option is to replace the copper tubing with Manguera flexible hoses that are more expensive but will result in a much lower failure rate of $\lambda_4 = 0.025$. Another option available is to replace the entire set of connectors by a single unit that will eliminate two connections and is much easier to install. The cost savings in units of $1000 for the alternatives under poor, moderate, and good sales conditions are given in the following decision matrix.

Alternatives	Sales Conditions, Θ:		
	Poor	Moderate	Good
Copper tube system	0	0	0
Manguera cable	−25	50	85
Single unit design	−35	70	100

Find the best design option given the probabilities for the poor, moderate, and good sales conditions are: $P(\theta_1) = 1/4$, $P(\theta_2) = 1/2$, and $P(\theta_3) = 1/4$.

References

ARINC Corporation, 1964. *Reliability Engineering*, W.H. Von Alven, Ed., Prentice Hall, Upper Saddle River, NJ.

Ebeling, C.E., 2005, *An Introduction to Reliability and Maintainability Engineering*, Waveland Press, Long Grove, IL.

Ivanovic, G., 2000, The reliability allocation application in vehicle design, *Int. J. Veh. Design*, 24, 274-286.

Jaynes, E.T., 1957, Information theory and statistical mechanics, *Phys. Rev.*, 106, 620-630.

Kuo, W., and M.J. Zuo, 2003, *Optimal Reliability Modeling: Principles and Applications*, John Wiley & Sons, New York.

Military Handbook: *Reliability Prediction of Electronic Equipment* (MIL-HDBK-217F), Rome Air Development Center, Griffiss, AFB, New York, 1991.

MIL-STD-1629, 1980, Procedures for Performing a Failure Modes, Effects and Criticality Analysis, U.S. Department of Defense, Washington, D.C.

Raiffa, H. 1970, *Decision Analysis*, Addison-Wesley, Reading, MA.
Roberts, H.R., W.E. Vesley, D.F. Haast, and F.F. Goldberg, 1981, *Fault Tree Handbook*, U.S. Nuclear Regulatory Commission, NUREG-0492.
Thomas, M.U. and J.P. Richard, 2006, Warranty-based methods for establishing reliability improvement targets, *IIE Transactions*, (to appeaar).
Thomas, M.U., 1979, A generalized maximum entropy principle, *Oper. Re.*, **27**, 1188–1196.
Tribus, M., 1969, *Rational Descriptions Decisions and Designs*, Pergamon Press, New York.

7
Toward an Integrated Product Quality System

7.1 Introduction

Product quality is the collective influence of several elements that affect how well a product is made, how the item performs, and how well customers accept it. Reliability has impact on all of these issues and, therefore, is perhaps the most critical of all of the quality elements. In Chapters 2 and 3, we discussed methods for analyzing and assessing product failures and reliability. The implications of reliability and quality relative to warranty policies are apparent in the warranty cost models developed in Chapter 4. When quality is high the resulting number on failures and associated costs for warranty will be low, whereas a low-quality product can give rise to high warranty servicing costs. This simple and essentially trite logic is the basis for using warranty claims statistics for constructing the relative quality indicators and MOP/MIS (month of production/month in service) charts for monitoring quality in Chapter 5. Warranty information and performance are very important in trying to control overall quality output to consumers and it also provides valuable feedback for diagnosing areas for seeking quality improvements.

A basic framework now exists for developing cost relationships for evaluating policies and economic tradeoffs involving reliability and quality decisions. In this chapter we will review the emergence of product quality as an important function of the success of manufacturing enterprises and present a paradigm for moving closer to an integrated product quality system. This paradigm is based on the threesome of quality, reliability, and warranty.

7.2 The Quality Movement

Product quality as we know it today is the result of an evolutionary process that has been shaped over time by the growth and development of our manufacturing enterprises. Our present state-of-the-art in manufacturing and quality management is the result of an enormous number of creations and developments in technology, economic developments, and social changes throughout the world. Manufacturing started from the early proprietor shopkeeper, who produced made-to-order items using simple hand tools and eventually using machines. By the early 1800s, machine developments, such as John Wilkinson's boring machine and the introduction of interchangeable parts by Eli Whitney, made important advances in production. The operations continued to be predominantly manual for many years, though the focus was on increasing production output. Quality was more at the good grace of the producer's pride and attitude than anything else since there was little to no competition. Customers essentially had the option of purchasing or not purchasing the items that were available on the market. The drive for high-speed production grew another notch at the turn of the 19th century with the introduction of the assembly line by Henry Ford and the scientific management methods by Fredrick Taylor. The manufacturing philosophy was to continue to maximize production output through effective use of labor and the infusion of automation. A detailed account of the historical perspective and development of automation can be found in Groover (2001).

Quality was primarily viewed as quality control methods for conforming to the process standards and requirements. This pattern of mass production at minimum cost without dedicated customer focus continued for many years until the mid-1970s when there was a drastic change in the world economy. For the first time ever, the U.S. economy faced extreme competition in the world markets from products marketed at not only lower prices but at higher quality. The period of 1975 to 1985 was a decade of a major transition in the concept of quality from "producer-interpreted quality" to "customer-interpreted quality," which has produced a massive impact on product manufacturing throughout the world.

7.2.1 Quality Control Philosophies

The interpretation of product quality and the approaches for achieving and maintaining high-quality products have been studied by scholars for many years. Among them are three pioneers: Walter A. Deming, Philip B. Crosby, and Joseph M. Juran. These three individuals are credited with providing the basic guidance that is used in the development of quality management programs throughout the world. A detailed discussion of their philosophies can be found in Mitra (1998). The essence of these works as they apply to product quality systems is examined below.

7.2.1.1 Deming Philosophy

In an ideal organization, management, workers, investors, and suppliers are a team. The workforce is divided into management and workers, and 85% of the quality problems that occur are resolvable only by management. In order to achieve high quality, a major cultural change is necessary for the entire organization. Planning should be for the long run, with planned courses of action for the short run. Deming proposes 14 points for management to employ, the most controversial being his position on establishing goals. He does not believe in setting numerical quotas and managing by objectives. He does advocate the use of statistical process control methods. (See Deming, 1982)

7.2.1.2 Crosby Philosophy

The process starts with a thorough evaluation of the existing quality system by applying a method that consists of a matrix for identifying operations in the production process with improvement potential. This method, called a quality management maturity grid, categorizes the capabilities of the management, organization, and relative costs of quality by the level of maturity in dealing with the issues. A 14-step quality improvement program is then developed and implemented. Crosby proposes a "zero defects per day" goal that is understood and supported by all employees. He is an advocate for the use of quality teams, goal setting, and rewards for improving quality. His measurement of quality improvement focuses on improving conformance quality by reducing scrap, rework, inventory, inspection and testing, and service. (See Crosby, 1979)

7.2.1.3 Juran Philosophy

Management adopts a unified approach to quality as defined by the fitness of a product for use by customers. Juran employs a trilogy process of (1) quality planning, (2) quality control, and (3) quality improvement. Planning is a critical element of the process and it starts by identifying the customers and their needs, both internal and external to the organization. Product features are then developed to meet these needs, followed by processes for achieving them. For the quality control element, measurements of standards and performance are selected as they relate to the customer requirements. A continuous process of improvement is employed for identifying and diagnosing problems and their causes and remedies. (See Juran and Gryna, 1993)

All of these philosophies have three things in common: (1) organizations should have an integrated total quality system, (2) top management must be fully committed to the program, and (3) the system should have a built-in continuous improvement element.

The difference between Deming, Crosby, and Juran is in their approaches to structuring improvement strategies. Deming's focus is on altering the culture of the organization to improve the processes, while Crosby and Juran focus on diagnosing and identifying improvement alternatives that will positively impact customers.

7.2.2 A Modern View of Quality

Quality was described in Chapter 3 as a multiple attribute vector of six components that relate to the way a product is designed, developed, and used by

TABLE 7.1

Product Quality Dimensions from Garvin (1987)

1. **Performance** — Operational performance of the product, such as vehicle fuel consumption in miles per gallon, heater power output in British Thermal Units (BTUs), and the picture resolution in pixels for a digital camera.
2. **Durability** — Ultimate amount of use before the product deteriorates or fails beyond repair.
3. **Reliability** — Probability of product failing within a specified time having survived to that point.
4. **Conformance** — Degree to which design and operating characteristics comply with preestablished standards.
5. **Aesthetics** — Way in which the product is actually sensed through appearance, feel, sound, touch, and smell.
6. **Perceived Quality** — Overall image of the product among users and potential users.
7. **Features** — Additional options and characteristics that distinguish a product from other competition.
8. **Serviceability** — Ease in servicing and maintaining a product.

consumers. David Garvin's original concept of quality included the attributes: features and serviceability (Garvin, 1987), which are listed in Table 7.1. Not only do these attributes describe quality, but they also provide reference to areas for seeking improvements. The philosophies of Deming, Crosby, and Juran are pertinent to this framework in a broad sense as well, but only in the case of Juran does the philosophy specifically address these quality dimensions. Deming's focus is on improving processes using statistical quality control methods. Crosby does focus on conformance quality, but Juran's view directly relates to providing customers with satisfaction by providing them with desirable product features. He also emphasizes performance and maintenance free service.

7.3 A QRW (Quality, Reliability, and Warranty) Paradigm

In Chapter 5, we discussed the relationships among quality, reliability, and warranty. Reliability being an extremely critical element of quality, and measurable by conventional methods is a natural element to target when seeking quality improvement alternatives. Reliability problems are also quite visible in warranty data. Warranty feedback is a primary source of information that manufacturers collect and use as quality indicators for tracking quality performance, such as the MOP/MIS charting methods. The next step in achieving a greater control and response for improving quality is to develop a more granular means for relating feedback information from warranty and other sources to the quality attributes of a product.

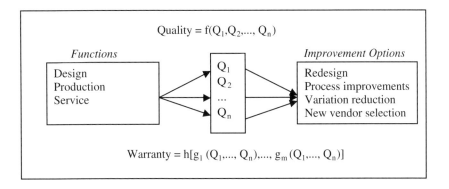

FIGURE 7.1
A QRW paradigm for quality improvement.

A QRW paradigm for quality improvement is illustrated in Figure 7.1. The organization is presumed to be completely committed to total quality and continuous improvement in concurrence with the philosophies of Crosby, Deming, and Juran. This is consistent with most manufacturers of consumer products and essential in being competitive in global markets. Quality is a function of prescribed attributes $Q_1, Q_2, ..., Q_n$, which could be the eight (or fewer) attributes by Garvin in Table 7.1, or even others that more directly characterize a particular product line. While the set of these n attributes collectively describe quality, their relative dependence on each of the processes in the design, production, and service of the product will of course vary. For example, the design of an item relates to all elements, but much more to reliability than the others. Conformance quality, on the other hand, is generally mostly influenced by variations within the production processes. Note also that each Q_i can and often does have several measures for defining its level of quality. The performance of an automobile, for example, is characterized by its horsepower, acceleration, braking distance, ride, etc.

The feedback from the warranty, $h[g_1(Q_1,...,Q_n),...,g_m(Q_1,...,Q_n)]$, is a function of m measurements about the quality vector $(Q_1,...,Q_n)$. This could be based on MOP/MIS data for warranty costs or number of warranty claims, or a related quantifiable variable. Ideally, each $g_j(Q_1,...,Q_n)$ has diagnostic qualities in allowing translations of the information back to the design, production, and service functions. The current methods that are used in practice have some capabilities for broadly directing to general areas as sources of failures and high warranty cost components. The quality of the data is often an issue as well as the degree of collection that is built into the database. With improved methods, the elements of $g_1,...,g_m$ could provide further direction for determining alternatives for making improvements in the product quality through any or all of redesign, specific process changes, methods for reducing process variation, and replacing vendors.

7.4 Some Concluding Remarks

It is highly unlikely that products manufactured and sold to consumers will ever achieve perfect quality because people by their nature will alter their preferences, demands, and expectations for products as technology advances and provides new options. The QRW paradigm of Figure 7.1 provides direction toward an integrated product system concept. The path to further improvement requires new technology in sensing, information technology methods for rapid access to data and information, and condition monitoring for processes that will allow for greater detail in detecting and diagnosing problems.

In this book, we have taken the view of the manufacturer who understands the importance of customers and their impact on quality. The next step in a path toward what might be considered ultra high quality requires even greater focus on the customer. New technologies are needed for further integration of equipment and processes without variation. Seamless feedback is necessary for all aspects of the quality vector. Inspections as we know them will be replaced by seamless and invisible elements of production operations. Errors and defects are avoided by early detection of leading indicators. Diagnostic response capabilities have to be extremely fast and integrated with online control and sensing that has built-in self-rectification capabilities. For example, a ceramic grinding operation has a feed-forward capability for detecting the onset of a crack formation, which automatically initiates adjustments that alter the machine speed to avoid the occurrence of the crack.

Unlike the early days of mass production, the entire world has learned and has since responded to the need for making quality a priority in product manufacturing. Since the shift from production-interpreted quality to a customer-accepted interpretation (during the period 1975 through 1985), industry has responded through several developments that have advanced manufacturing with quality as an integral part of the system. This is well exemplified in the current lean-production concept that adapts the essence of the accepted quality philosophies with effective production and manufacturing methods to achieve ultimate manufacturing excellence.

References

Crosby, P.B., 1979, *Quality is Free*, McGraw-Hill, New York.
Deming, W.E., 1982, *Quality, Productivity, and Competitive Position*, Cambridge, MA: Center for Advanced Engineering Study, Massachusetts Institute of Technology.
Garvin, D.A., 1987, Competing on the eight dimensions of quality, *Harvard Business Review*, vol. 65, No. 6, pp. 107–109.

Groover, M.P., 2001, *Automation, Production Systems, and Computer-Integrated Manufacturing*, 2nd ed., Prentice Hall, Upper Saddle River, NJ.

Juran, J.M. and F.M. Gryna, 1993, *Quality Planning and Analysis*, 3rd ed., McGraw-Hill, New York.

Mitra, A., 1998, *Fundamentals of Quality Control and Improvement*, 2nd ed., Prentice-Hall, Upper Saddle River, NJ, Chap. 2.

Appendix A

Notations and Acronyms

Chapter 1

AGREE	Advisory Group on Reliability of Electronics Equipment
cdf	cumulative distribution function
CFR	constant failure rate
CPUS	warranty cost per unit sold
DFR	decreasing failure rate
dpf	discrete probability function
DPUS	defects per unit sold
FMEA	failure mode and effects analysis
FMECA	failure mode, effect, and criticality analysis
FRW	free replacement warranty policy
FTA	failure tree analysis
IFR	increasing failure rate
MIL-STD	military standard
MIS	month in service
MOP	month of production
MTTF	mean time to failure
pdf	probability density function
PRW	prorate replacement warranty
TPL	total product life
TQS	total quality system

$E[.]$	expected value operator
$E[X^k]$	k^{th} moment about 0
$f(t)$	probability density function for failure times
$F(t)$	cumulative failure time distribution function
$\Gamma(x)$	gamma function
m	mean; $E[X]$
$\bar{x}$	sample mean
$R(t)$	reliability function
$R_i(t)$	reliability function for component i
$R_S(t)$	system reliability
s^2, s	sample variance, standard deviation
Var[X]	variance of X

Chapter 2

$\bar{f}(s)$	Laplace transform of pdf f(.)
$f_1 * f_2$	convolution of pdf's f_1 and f_2
β, θ	slope and characteristic value parameters for the Weibull distribution
χ^2	chi-square test statistic
$\chi^2_{\alpha;\nu}$	value from the χ^2 distribution tables for level α and ν degrees of freedom
$E[X^{(m)}]$	m^{th} factorial moment
e_j	expected number of events in cell j
f_j	number of observations in cell j
$\hat{F}$	fitted distribution from observations
F(x;p)	distribution of random variable X with p parameters
M(t)	renewal function
$M_A(t)$	asymptotic approximation to M(t)
$M_L(t)$	Leadbetter approximation to M(t)
N(t)	number of events in (0,t)
P(z)	probability generating function

Chapter 3

C	a random variable representing capacity
G(u,u+t)	average failure rate over (u,u+t)
I	interference random variable between load and capacity
L	a random variable representing load
$\lambda(t)$	hazard function
p	interference probability of a failure
T	nonnegative random variable representing failure times

Appendix A

Chapter 4

	$B(w)$	expected cost benefit for warranty over $(0,w)$
	$C(w)$	expected warranty cost
	c_0	initial unit cost to the manufacturer for a warranty event
	c_1	unit repair/replacement cost to the manufacturer
	d	discount rate
	K	fixed cost to the manufacturer for a warranty program
	$m(t)$	renewal density function
	p	unit price
	π	expected price change per period
	$R(w)$	reserve resources for warranty of length $w > 0$
	θ	rate of return
	$X(t)$	unit cost to the manufacturer at time t
	$Y(w)$	manufacturer's total warranty cost at $w > 0$

Chapter 5

a_j	relative importance factor for element j
α_k	probability of a k-month lag in reporting
$C(i,j)$	expected warranty cost for items produced in month i with j months in service
$D_j(t)$	average number of defects per unit sold during MIS-j
$g(\pi)$	assessment function for $\pi(Q)$
L	random variable representing reporting lag
m_{ij}	number of claims from month of production i during month in service j
$\tilde{M}(a,b)$	estimated number of claims during (a,b)
$\hat{\lambda}_t$	maximum likelihood estimator for λ_t
$\tilde{\Lambda}_t$	predicted average number of claims per vehicle
$n_{x,t,r}$	number of claims at age t placed in service on day x with a lag of r days
N_i	accumulation of claims from production month i
N_{ij}	number of claims from production month i that occurred during MIS-j
$N_i(k)$	accumulation of claims from MOP-i up to month k
$N_{ijl L=k}$	number of claims generated from MOP-i during MIS-j with reporting lag of k months
$NR(t)$	number of reported claims at month t
$\pi_j(Q_j)$	quality performance measure for element j
$\pi(Q)$	overall quality performance
Q	quality vector
Q_i	quality element i
X_{ijk}	cost for claim k from production month i during MIS-j
z_{ij}	warranty cost for units produced in month i with j months in service
$Z_j(t)$	average warranty cost per unit sold during MIS-j

Chapter 6

α_{jk}, β_{jk}	known bounds on $\xi_k(x_i)$
A	set of decision alternatives $a_1, a_2, ..., a_m$
β_i	warranty burden for component i
c_{wi}	manufacturer's unit warranty cost for component i
$C(a_i, \theta_j)$	cost for alternative a_i with future outcome θ_j
$\xi_k(x_i)$	function k representing known constraints and conditions on a random variable X
$EAC(a_i)$	equivalent annual cost for alternative a_i
γ	reliability improvement goal
$H(p)$	entropy for discrete probability function p
λ^0	improved system failure rate
λ_i^0	improved failure rate for component i
$M_j(t)$	expected number of claims for component j during (0,t)
$M_j^0(T_w)$	expected number of claims for component j during warranty following improvement
p_i	probability of event i occurring
$R_i^0(T_w)$	reliability of component i during warranty following improvement
T_w	length of warranty
Θ	set of future outcomes of nature $\theta_1, \theta_2, ..., \theta_n$
V	set of consequences for $\{(a_i, \theta_j) : i = 1, ..., m; j = 1, ..., n\}$
v_{ij}	outcome values for consequences $\{(a_i, \theta_j) : i = 1, ..., m; j = 1, ..., n\}$

Appendix B

Selected Tables

TABLE B.1

Standard Normal Cumulative Probabilities

z	0.00	0.01	0.02	0.03	0.04	0.05	0.06	0.07	0.08	0.09
0.0	0.5000	0.5040	0.5080	0.5120	0.5160	0.5199	0.5239	0.5279	0.5319	0.5359
0.1	0.5398	0.5438	0.5478	0.5517	0.5557	0.5596	0.5636	0.5675	0.5714	0.5753
0.2	0.5793	0.5832	0.5871	0.5910	0.5948	0.5987	0.6026	0.6064	0.6103	0.6141
0.3	0.6179	0.6217	0.6255	0.6293	0.6331	0.6368	0.6406	0.6443	0.6480	0.6517
0.4	0.6554	0.6591	0.6628	0.6664	0.6700	0.6736	0.6772	0.6808	0.6844	0.6879
0.5	0.6915	0.6950	0.6985	0.7019	0.7054	0.7088	0.7123	0.7157	0.7190	0.7224
0.6	0.7257	0.7291	0.7324	0.7357	0.7389	0.7422	0.7454	0.7486	0.7517	0.7549
0.7	0.7580	0.7611	0.7642	0.7673	0.7704	0.7734	0.7764	0.7794	0.7823	0.7852
0.8	0.7881	0.7910	0.7939	0.7967	0.7995	0.8023	0.8051	0.8079	0.8106	0.8133
0.9	0.8159	0.8186	0.8212	0.8238	0.8264	0.8289	0.8315	0.8340	0.8365	0.8389
1.0	0.8413	0.8438	0.8461	0.8485	0.8508	0.8531	0.8554	0.8577	0.8599	0.86214
1.1	0.8643	0.8665	0.8686	0.8708	0.8728	0.8749	0.8770	0.8790	0.8810	0.8830
1.2	0.8849	0.8869	0.8888	0.8907	0.8925	0.8944	0.8962	0.8980	0.8997	0.9015
1.3	0.9032	0.9049	0.9066	0.9082	0.9099	0.9115	0.9131	0.9147	0.9162	0.9177
1.4	0.9192	0.9207	0.9222	0.9236	0.9251	0.9265	0.9279	0.9292	0.9306	0.9319
1.5	0.9332	0.9345	0.9357	0.9370	0.9382	0.9394	0.9406	0.9418	0.9429	0.9441
1.6	0.9452	0.9463	0.9474	0.9484	0.9495	0.9505	0.9515	0.9525	0.9535	0.9545
1.7	0.9554	0.9564	0.9573	0.9582	0.9591	0.9599	0.9608	0.9616	0.9625	0.9633
1.8	0.9641	0.9648	0.9656	0.9664	0.9671	0.9678	0.9686	0.9693	0.9699	0.9706
1.9	0.9712	0.9719	0.9726	0.9732	0.9738	0.9744	0.9750	0.9756	0.9761	0.9767
2.0	0.9773	0.9778	0.9783	0.9788	0.9793	0.9798	0.9803	0.9808	0.9812	0.9817
2.1	0.9821	0.9826	0.9830	0.9834	0.9838	0.9842	0.9846	0.9850	0.9854	0.9857
2.2	0.9861	0.9864	0.9868	0.9871	0.9875	0.9878	0.9881	0.9884	0.9887	0.9890
2.3	0.9893	0.9896	0.9898	0.9901	0.9904	0.9906	0.9909	0.9911	0.9913	0.9916
2.4	0.9918	0.9920	0.9922	0.9925	0.9927	0.9929	0.9931	0.9932	0.9934	0.9936
2.5	0.9938	0.9940	0.9941	0.9943	0.9945	0.9946	0.9948	0.9949	0.9951	0.9952
2.6	0.9953	0.9955	0.9956	0.9957	0.9959	0.9960	0.9961	0.9962	0.9963	0.9964
2.7	0.9965	0.9966	0.9967	0.9968	0.9969	0.9970	0.9971	0.9972	0.9973	0.9974
2.8	0.9974	0.9975	0.9976	0.9977	0.9977	0.9978	0.9979	0.9979	0.9980	0.9981
2.9	0.9981	0.9982	0.9983	0.9983	0.9984	0.9984	0.9985	0.9985	0.9986	0.9986
3.0	0.9987	0.9987	0.9987	0.9988	0.9988	0.9989	0.9989	0.9989	0.9990	0.9990
3.1	0.9990	0.9991	0.9991	0.9991	0.9992	0.9992	0.9992	0.9992	0.9993	0.9993
3.2	0.9993	0.9993	0.9994	0.9994	0.9994	0.9994	0.9994	0.9995	0.9995	0.9995
3.3	0.9995	0.9995	0.9996	0.9996	0.9996	0.9996	0.9996	0.9996	0.9996	0.9997
3.4	0.9997	0.9997	0.9997	0.9997	0.9997	0.9997	0.9997	0.9997	0.9997	0.9998

TABLE B.2
Chi-Square Distribution

χ^2 Table

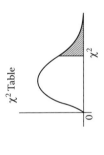

Degrees of freedom	P = 0.99	0.98	0.95	0.90	0.80	0.70	0.50	0.30	0.20	0.10	0.05	0.02	0.01
1	0.000157	0.000628	0.00393	0.0158	0.0642	0.148	0.455	1.074	1.642	2.706	3.841	5.412	6.635
2	0.0201	0.0404	0.103	0.211	0.446	0.713	1.386	2.408	3.219	4.605	5.991	7.824	9.210
3	0.115	0.185	0.352	0.584	1.005	1.424	2.366	3.665	4.642	6.251	7.815	9.837	11.341
4	0.297	0.429	0.711	1.064	1.649	2.195	3.357	4.878	5.989	7.779	9.488	11.668	13.277
5	0.554	0.752	1.145	1.610	2.343	3.000	4.351	6.064	7.289	9.236	11.070	13.388	15.086
6	0.872	1.134	1.635	2.204	3.070	3.828	5.348	7.231	8.558	10.645	12.592	15.033	16.812
7	1.239	1.564	2.167	2.833	3.822	4.671	6.346	8.383	9.803	12.017	14.067	16.622	18.475
8	1.646	2.032	2.733	3.490	4.594	5.527	7.344	9.524	11.030	13.362	15.507	18.168	20.090
9	2.088	2.532	3.325	4.168	5.380	6.393	8.343	10.656	12.242	14.684	16.919	19.679	21.666
10	2.558	3.059	3.940	4.865	6.179	7.267	9.342	11.781	13.442	15.987	18.307	21.131	23.209
11	3.053	3.609	4.575	5.578	6.989	8.148	10.341	12.899	14.631	17.275	19.675	22.618	24.725
12	3.571	4.178	5.226	6.304	7.807	9.034	11.340	14.011	15.812	18.549	21.026	24.054	26.217
13	4.107	4.765	5.892	7.042	8.634	9.926	12.340	15.119	16.985	19.812	22.362	25.472	27.688
14	4.660	5.368	6.571	7.790	9.467	10.821	13.339	16.222	18.151	21.064	23.685	26.873	29.141
15	5.229	5.985	7.261	8.547	10.307	11.721	14.339	17.322	19.311	22.307	24.996	28.259	30.578
16	5.812	6.614	7.962	9.312	11.152	12.624	15.338	18.418	20.465	23.542	26.296	29.633	32.000
17	6.408	7.255	8.672	10.085	12.002	13.531	16.338	19.511	21.615	24.769	27.587	30.995	33.409

Appendix B

n													
18	7.015	7.906	9.390	10.865	12.857	14.440	17.338	20.601	22.760	25.989	28.869	32.346	34.805
19	7.633	8.567	10.117	11.651	13.716	15.352	18.338	21.689	23.900	27.204	30.144	33.687	36.191
20	8.260	9.237	10.851	12.443	14.578	16.266	19.337	22.775	25.038	28.412	31.410	35.020	37.566
21	8.897	9.915	11.591	13.240	15.445	17.182	20.337	23.858	26.171	29.615	32.671	36.343	38.932
22	9.542	10.600	12.338	14.041	16.314	18.101	21.337	24.939	27.301	30.813	33.924	37.659	40.289
23	10.196	11.293	13.091	14.848	17.187	19.021	22.337	26.018	28.429	32.007	35.172	38.968	41.638
24	10.856	11.992	13.848	15.659	18.062	19.943	23.337	27.096	29.553	33.196	36.415	40.270	42.980
25	11.524	12.697	14.611	16.473	18.940	20.867	24.337	28.172	30.675	34.382	37.652	41.566	44.314
26	12.198	13.409	15.379	17.292	19.820	21.792	25.336	29.246	31.795	35.563	38.885	42.856	45.642
27	12.879	14.125	16.151	18.114	20.703	22.719	26.336	30.319	32.912	36.741	40.113	44.140	46.963
28	13.565	14.847	16.928	18.939	21.588	23.647	27.336	31.391	34.027	37.916	41.337	45.419	48.278
29	14.256	15.574	17.708	19.768	22.475	24.577	28.336	32.461	35.139	39.087	42.557	46.693	49.588
30	14.953	16.306	18.493	20.599	23.364	25.508	29.336	33.530	36.250	40.256	43.773	47.962	50.892

For degrees of freedom greater than 30, the expression $\sqrt{2x^2} - \sqrt{2n' - 1}$ may be used as a normal deviate with unit variance, where n' is the number of degrees freedom.

Reproduced from *Statistical Methods for Research Workers, 6th ed.*, with the permission of the author, R. A. Fisher, and his publisher, Oliver and Boyd, Edinburgh.

TABLE B.3

Laplace Transforms

$f(x)$	$\psi(\theta) = \int_0^\infty e^{-\theta x} f(x) dx$
$af(x) + bg(x)$	$a\psi_f(\theta) + b\psi_g(\theta)$
$\dfrac{d}{dx} f(x)$	$\theta f(\theta) - \psi(+0)$
$\dfrac{d^n}{dx^n} f(x)$	$\theta^n \psi(\theta) - \theta^{n-1} \psi(+0) - \theta^{n-2} \dfrac{d}{d\theta} \psi(+0) - \ldots - \dfrac{d^n}{d\theta^n} \psi(+0)$
$\int_0^x f(u) du$	$\dfrac{1}{\theta} \psi(\theta)$
$\int_0^x \int_0^y f(u) du dv$	$\dfrac{1}{\theta^2} \psi(\theta)$
$\int_0^x f(y) g(x-y) dy$	$\psi_f(\theta) \psi_g(\theta)$
$e^{ax} f(x)$	$\psi(\theta - a)$
$f(x-a) u(x-a)^1$	$e^{-a\theta} \psi(\theta)$
$x^n f(x)$	$(-1)^n \dfrac{d^n}{d\theta^n} \psi(\theta)$
$\dfrac{1}{x} f(x)$	$\int_\theta^\infty \psi(\theta) d\theta$
$f(ax),\ a > 0$	$\dfrac{1}{a} \psi(\theta/a)$
$\lim_{x \to 0} f(x)$	$\lim_{\theta \to \infty} \theta \psi(\theta)$
$\lim_{x \to \infty} f(x)$	$\lim_{\theta \to 0} \theta f(\theta)$
$u(x)$	1
c	$\dfrac{c}{\theta}$
$x^n,\ n = 1, 2, \ldots$	$\dfrac{n!}{\theta^{n+1}}$
$x^a,\ a > 0$	$\dfrac{\Gamma(a+1)}{\theta^{a+1}}$
e^{ax}	$\dfrac{1}{\theta - a}$

(continued)

TABLE B.3
Laplace Transforms (Continued)

xe^{ax}	$\dfrac{1}{(\theta-a)^2}$
$\dfrac{x^{n-1}e^{ax}}{\Gamma(n)}$	$\dfrac{1}{(\theta-a)^n}$
$u(x-a)$	$\dfrac{e^{-a\theta}}{\theta}$
$\sin(ax)$	$\dfrac{a}{\theta^2+a^2}$
$\cos(ax)$	$\dfrac{\theta}{\theta^2+a^2}$

[1] Heaviside or "unit step" function $u(x-1) = \begin{cases} 1, & x > a \\ 0, & x \le a \end{cases}$

TABLE B.4

Probability Generating Functions Operations

$p(x)$	$P_x(z) = \sum_{x=0}^{\infty} zx p(x)$
$ap(x) + bq(y)$	$aP_x(z) + bQ_y(z)$
$\Delta p(x) = p(x+1) - p(x)$	$\left(\dfrac{1}{z} - 1\right) P_x(z) - \dfrac{1}{z} P_x(0)$
$p(x + k)$	$z^{-k}\left(P_x(z) - \sum_{i=0}^{k-1} z^i p(i)\right)$
$\sum_{i=0}^{x} p(i)$	$\dfrac{1}{1-z} P_x(z)$
$\sum_{i=0}^{x} p(i) q(x-i)$	$P(z)Q(z)$
$k^{rx} p(x)$	$P_x(k^r z)$
$p(x-k) u(x-k)^1$	$z^k P_x(z)$
$\lim_{x \to 0} p(x)$	$P_x(0^+)$
$\lim_{x \to \infty} p(x)$	$\lim_{z \to 1}(1-z) P_x(z)$
c	$\dfrac{c}{1-z}$
c^x	$\dfrac{1}{1-cz}$
x^k	$\left(z \dfrac{d}{dz}\right)^k \dfrac{1}{1-z}$
$\dfrac{c^x}{x!}$	e^{cz}
xc^x	$\dfrac{cz}{(1-cz)^2}$
$x^2 c^x$	$\dfrac{cz(1+cz)}{(1-cz)^3}$
$\dbinom{x+k}{k} c^x$	$\dfrac{1}{(1-cz)^{k+1}}$
$\dbinom{a}{x} b^x c^{a-x}$	$(c+bz)^a$

(continued)

TABLE B.4

Probability Generating Functions Operations (Continued)

$\sin(cx)$	$\dfrac{z \sin c}{z^2 + 1 - 2z \cos c}$
$\cos(cx)$	$\dfrac{1 - z \cos c}{z^2 + 1 - 2z \cos c}$

[1] Heaviside or "unit step" function $u(x-k) = \begin{cases} 1, & x > k \\ 0, & x \leq k \end{cases}$

Appendix C

Counting Distributions

C.1 Binomial Counting Distribution

Let $Y_1, Y_2, \ldots$ be a sequence of inspection outcomes represented by independent Bernoulli trials with failure event probability $0 < p < 1$. The number of inspections, X_1 until a failure occurs, therefore, is distributed geometric with

$$p_{X_1}(x) = pq^{x-1}, \quad q = 1-p; x = 1,2,\ldots \qquad (C.1.1)$$

Proposition 1

The number of inspections, S_r, until the rth failure occurs is distributed negative binomial with

$$g_{S_r}(y) = \binom{y-1}{r-1} p^r q^{y-r}, \quad y = r, r+1, \ldots \qquad (C.1.2)$$

Proof

It follows that

$$S_r = X_1 + \cdots + X_r$$

where $X_1, X_2, \ldots, X_r$ are independent and identically distributed geometric with discrete probability function Equation (C.1.1) and associated probability generating function,

$$P(z) = \frac{pz}{1-qz} \qquad (C.1.3)$$

The probability generating function of the discrete probability function for S_r is therefore,

$$G_{S_r}(z) = \prod_{i=1}^{r} P_{X_i}(z) = \left(\frac{pz}{1-qz}\right)^r \qquad (C.1.4)$$

Inversion, using Table B.4 above, shows that S_r has the discrete probability function given in (C.1.2).

Proposition 2

The number of failures, N, that occur in (0,n) inspections is distributed binomial with discrete probability function

$$g_N(x) = \binom{n}{x} p^x q^{n-x}, \quad x = 0, 1, \ldots, n \tag{C.1.5}$$

Proof

Since the inspection outcomes are independent Bernoulli trials with

$$p_Y(y) = P(Y_1 = 1) = p^y q^{1-y}; \quad y = 0, 1 \tag{C.1.6}$$

then the number of failure events is the sum

$$N = Y_1 + \cdots + Y_n$$

The probability generating function for $p_Y(y)$ from Table B.4 is

$$P_Y(z) = q + pz \tag{C.1.7}$$

and, hence, for $g_N(y)$ we have

$$G_N(z) = \prod_{i=1}^{n} P_{Y_i}(z) = [P_{Y_1}(z)]^n$$

or

$$G_N(z) = (q + pz)^n \tag{C.1.8}$$

which is the generation function for the binomial.

C.2 Poisson Counting Distribution

Let $X_1, X_2, \ldots$ be random variables representing failure times that are independent and identically distributed exponential with

$$f(x) = \lambda e^{-\lambda x}, \quad x \geq 0 \tag{C.2.1}$$

Proposition 3

The distribution of the time to the rth failure is distributed Erlang with

$$g_r(y) = \frac{\lambda(\lambda y)^{r-1}}{(r-1)!} e^{-\lambda y}; \quad r = 1, 2, \ldots; \; \lambda \geq 0 \tag{C.2.2}$$

Proof

The time to the rth failure is the sum

$$S_r = X_1 + \cdots + X_r$$

Since these random variables are independent, it follows that the Laplace transform for the probability density function for $g_{S_r}(y)$ is

$$\bar{g}(s) = E(e^{-s(X_1+\cdots+X_r)}) = \prod_{i=1}^{r} \left(\bar{f}_{X_i}(s)\right) = \left(\bar{f}_X(s)\right)^r$$

For X_i distributed exponential, then $f_X(x)$ has the Laplace transform

$$\bar{f}_X(s) = \frac{\lambda}{\lambda + s} \tag{C.2.3}$$

therefore,

$$\bar{g}(s) = \left(\frac{\lambda}{\lambda + s}\right)^r \tag{C.2.4}$$

which can be verified in Table B.3 as the Laplace transform for the Erlang probability density function given in Equation (C.2.2).

Proposition 4

The number of failure events occurring in (0,t) is distributed Poisson with

$$p_{N(t)}(n) = \frac{(\lambda t)^n}{n!} e^{-\lambda t}; \quad n = 0, 1, \ldots; \; t \geq 0 \tag{C.2.5}$$

Proof

Note that in order for the event $\{N(t) \leq n\}$ to occur, it is necessary that $\{S_{n+1} > t\}$, so from (C.2.2) for n = 0

$$P\{N(t) \leq 0\} = P\{S_1 > t\} = \int_{t}^{\infty} \lambda e^{-\lambda \tau} d\tau = e^{-\lambda t} \tag{C.2.6}$$

For $n \geq 1$,

$$P\{N(t) \leq n\} = P\{S_{n+1} > t\} = \int_t^\infty \frac{\lambda(\lambda\tau)^n}{n!} e^{-\lambda\tau} d\tau \qquad \text{(C.2.7)}$$

Integrating by parts, let

$$u = \frac{\lambda(\lambda\tau)^n}{n!} \quad \text{and} \quad dv = e^{-\lambda\tau} d\tau$$

hence,

$$du = \frac{\lambda^2(\lambda\tau)^{n-1}}{(n-1)!} d\tau \quad \text{and} \quad v = \frac{e^{-\lambda\tau}}{-\lambda}$$

and it follows that

$$P\{N(t) \leq n\} = \frac{(\lambda t)^n}{n!} e^{-\lambda t} + \int_t^\infty \frac{\lambda(\lambda\tau)^{n-1}}{(n-1)!} e^{-\lambda\tau} d\tau \qquad \text{(C.2.8)}$$

The right-hand term in Equation (C.2.8) is $P\{N(t) \leq n-1\}$, therefore,

$$P\{N(t) \leq n\} = \frac{(\lambda t)^n}{n!} e^{-\lambda t} + P\{N(t) \leq n-1\}$$

from which it follows that

$$P\{N(t) = n\} = P\{N(t) \leq n\} - P\{N(t) \leq n-1\} = \frac{(\lambda t)^n}{n!} e^{-\lambda t}$$

C.3 Renewal Function for Erlang (λ,2) Distributed Times

Applying the transform result for M(t) from (2.52), in Chapter 2

$$\bar{M}(s) = \frac{\bar{f}(s)}{s[1 - \bar{f}(s)]}$$

Appendix C

The probability density function for the Erlang $(\lambda,2)$,

$$f(t) = \lambda t e^{-\lambda t}, \quad t \geq 0$$

has the Laplace transform

$$\bar{f}(s) = \left(\frac{\lambda}{\lambda+s}\right)^2$$

Substituting into $\bar{M}(s)$,

$$\bar{M}(s) = \frac{\lambda^2}{s[(\lambda+s)^2 - \lambda^2]} = \frac{\lambda^2}{s^2(s+2\lambda)}$$

Expanding the right side by partial fractions,

$$\frac{\lambda^2}{s[(\lambda+s)^2 - \lambda^2]} = \frac{A_0}{s^2} + \frac{A_1}{s} + \frac{B}{(s+2\lambda)} \quad \text{(C.3.1)}$$

To determine the constants A_0 and A_1, we multiply the left side by s^2 obtaining

$$G(s) = \frac{\lambda^2}{s+2\lambda}, \quad G'(s) = \frac{-\lambda^2}{(s+2\lambda)^2}$$

and

$$A_0 = G(0) = \frac{\lambda}{2}$$

$$A_1 = G'(0) = -\frac{1}{4}$$

For B, it follows that $B = \lim_{s \to -2\lambda}\left(\frac{\lambda^2}{s^2}\right) = \frac{1}{4}$

Substituting these constants back into Equation (C.3.1),

$$\bar{M}(s) = \frac{\lambda/2}{s^2} - \frac{1/4}{s} + \frac{1/4}{s+2\lambda}$$

which can be inverted from tables to

$$M(t) = \frac{\lambda t}{2} - \frac{1}{4} + \frac{1}{4}e^{-2\lambda t}, \quad t \geq 0 \quad \text{(C.3.2)}$$

Appendix D

Solutions to Exercises

Chapter 2

1. $\bar{g}(s) = \int_0^\infty e^{-st} \frac{\lambda(\lambda t)^{r-1}}{(r-1)!} e^{-\lambda t} dt = \int_0^\infty \frac{\lambda(\lambda t)^{r-1}}{(r-1)!} e^{-(\lambda+s)t} dt$

   $y = (\lambda + s)t, \quad dt = dy/(\lambda+s)$

   $\bar{g}(s) = \frac{\lambda^r}{(r-1)!} \left(\frac{1}{\lambda+s}\right)^r \int_0^\infty y^{r-1} e^{-y} dy = \frac{\lambda^r}{(r-1)!} \left(\frac{1}{\lambda+s}\right)^r \Gamma(r)$

   since $\Gamma(r) = (r-1)!$

   $$\bar{g}(s) = \left(\frac{\lambda}{\lambda+s}\right)^r$$

   $\bar{g}'(s) = \frac{-r}{\lambda}, \quad \therefore E[T] = r/\lambda$

   $\bar{g}''(s) = \frac{r(r+1)}{\lambda^2} = E[T^2]$

   $\therefore \text{Var}\{T\} = r/\lambda^2$

2. $\bar{f}(s) = \int_{-\infty}^\infty e^{-st} \frac{e^{-\frac{1}{2}\left(\frac{t-\mu}{\sigma}\right)^2}}{\sigma\sqrt{2\pi}} dt = \int_{-\infty}^\infty \frac{e^{-\frac{1}{2}\left(\frac{t-\mu}{\sigma}\right)^2 - st}}{\sigma\sqrt{2\pi}} dt = \int_{-\infty}^\infty \frac{e^{-\left[\frac{(t-\mu)^2+2s\sigma^2 t}{2\sigma^2}\right]}}{\sigma\sqrt{2\pi}} dt$

   Completing the square in the numerator of the exponent term and simplyfing,

   $\bar{f}(s) = e^{-\mu s + \sigma^2 s^2/2} \int_{-\infty}^\infty \frac{e^{-\frac{1}{2}\left(\frac{t-(\mu-s\sigma^2)}{\sigma}\right)^2}}{\sigma\sqrt{2\pi}} dt = e^{-\mu s + \sigma^2 s^2/2}$

since the integral term is equal to 1.

3. $\bar{f}_Y(s) = \prod_1^n \bar{f}_{X_i}(s) = (e^{-\mu s + \sigma^2 s^2/2})^n = e^{-n\mu s + n\sigma^2 s^2/2}$

4. $Y = X_1 + X_2$

$$f(x_1) = f(x_2) = \begin{cases} 1/a, & 0 < x \le a \\ 0, & x > a \end{cases}$$

$$f_Y(y) = \int_0^y f_1(x_1) f_2(y - x_1) dx_1 = \int_0^y \frac{f_2(y - x_1)}{a} dx_1$$

$$= \begin{cases} y, & 0 \le y \le a \\ 2 - y, & 1 < y \le 2a \\ 0, & \text{otherwise} \end{cases}$$

5. $P(z) = \sum_0^\infty z^x p_X(x) = \frac{1}{8} + \frac{1}{4}z + \frac{1}{2}z^2 + \frac{1}{8}z^3$

a. $P(X \le 2) = P(X = 0) + P(X = 1) + P(X = 2)$

$P(X = 0) = p_X(0) = \frac{1}{8}$

$P(X = 1) = P'(0) = \left[\frac{1}{4} + z + \frac{3}{8}z^2\right]_{z \to 0} = \frac{1}{4}$

$P(X = 2) = \frac{1}{2}P''(0) = \frac{1}{2}\left[1 + \frac{6}{8}z^2\right]_{z \to 0} = \frac{1}{2}$

$\therefore P(X \le 2) = \frac{7}{8}$

$E[X] = P'(1) = \left[\frac{1}{4} + z + \frac{3}{8}z^2\right]_{z \to 1} = \frac{13}{8}$

$E[X(X - 1)] = P''(1) = \left[1 + \frac{6}{8}z^2\right]_{z \to 1} = \frac{14}{8}$

$\therefore E[X^2] = E[X(X - 1)] + E[X] = \frac{27}{8}$

Appendix D

$$Var[X] = \frac{27}{8} - \left(\frac{13}{8}\right)^2 = \frac{47}{64}$$

b. $Y = X_1 + X_2$

$$G_Y(z) = P_{X_1}(z)P_{X_2}(z) = \left(\frac{1}{8} + \frac{1}{4}z + \frac{1}{2}z^2 + \frac{1}{8}z^3\right)^2$$

$$G_Y'(z) = P_{X_1}'(z)P_{X_2}(z) + P_{X_1}(z)P_{X_2}'(z) = G_Y(z) = 2P_X'(z)P_X(z)$$

$$G_Y''(z) = 2\{P_X''(z)P_X(z) + (P_X'(z))^2\}$$

$$G_Y'(1) = 2P_X'(1)P_X(1) = 2P_X'(1) = 2E[X] = \frac{13}{4}$$

$$G_Y''(1) = 2\{P_X''(1)P_X(1) + (P_X'(1))^2\} = 2\left[\frac{14}{8} + \frac{13}{8}\right] = \frac{15}{2}$$

$$E[Y] = \frac{13}{4}, \quad Var[Y] = G_Y''(1) + G_Y'(1) - (G_Y'(1))^2 = \frac{47}{16}$$

7. Goodness of fit test

Interval	f_j	e_j	$(f_j - e_j)^2/e_j$
0 – 5	10	7.9	.5426
5 – 10	7	5.8	.4130
10 – 20	6	7.4	.2796
>20	7	8.9	.2201
			$\chi^2 = 1.4554$

Do not reject since $\chi^2_{.05;2} = 5.99 > 1.4554$

8. For $S_2 = X_1 + X_2$,

$$g_{S_2}(y) = \int_0^y \lambda^2 e^{-\lambda x} e^{-\lambda(y-x)} dx = \lambda^2 e^{-\lambda y} \int_0^y dx = \lambda^2 y e^{-\lambda y}$$

For $S_3 = X_3 + S_2$

$$g_{S_3}(y) = \int_0^y \lambda^3 x e^{-\lambda x} e^{-\lambda(y-x)} dx = \lambda^3 y e^{-\lambda y} \int_0^y dx = \frac{\lambda^3 y^2}{2} e^{-\lambda y}$$

. . .

$$g_{S_r}(y) = \frac{\lambda(\lambda y)^{r-1}}{(r-1)!} e^{-\lambda y}$$

9. $S_r = X_1 + X_2 + \cdots + X_r$, $\bar{f}_X(s) = \dfrac{\lambda}{\lambda + s}$

$$\bar{g}_{S_r}(s) = \prod_1^r \bar{f}_X(s) = \left(\dfrac{\lambda}{\lambda + s}\right)^r$$

which is the LT of the pdf,

$$g(y) = \dfrac{\lambda(\lambda y)^{r-1}}{(r-1)!} e^{-\lambda t}, \quad \lambda \geq 0, r = 1,2,\ldots, t \geq 0$$

10. $P[N_B(t) = x | N(t) = n] = \binom{n}{x}(1-\gamma)^x \gamma^{n-x}, \quad x = 0,1,\ldots,n$

$$P[N_B(t) = x] = \sum_0^\infty P[N_B(t) = x | N(t) = n] \dfrac{(\lambda t)^n}{n!} e^{-\lambda t}$$

$$= \sum_0^\infty \binom{n}{x}(1-\gamma)^x \gamma^{n-x} \dfrac{(\lambda t)^n}{n!} e^{-\lambda t} = \dfrac{(1-\gamma)^x}{x!} e^{-\lambda t} \sum_{n=x}^\infty \dfrac{\gamma^{n-x}(\lambda t)^{n-x}}{(n-x)!} e^{-\lambda t}$$

Since the summation term converges to $e^{\gamma \lambda t}$, it follows that

$$P[N_B(t) = x] = \dfrac{[(1-\gamma)\lambda t]^x}{x!} e^{-\lambda(1-\gamma)t}$$

11. a. $M_L(t) = \dfrac{1}{4}t^2 - \dfrac{1}{48}t^4 + \dfrac{5}{2304}t^6$;

$M_L(1/2) = 0.01696; \quad M_L(1) = 0.23134$

b. $M_A(t) = \dfrac{t}{2\sqrt{\pi}} + \dfrac{4-\pi}{\pi}$; $M_A(1/2) = 0.4143; \quad M_A(1) = 0.5553$

Chapter 3

1. $R(t_R) = R^*$
   a. Normal

Appendix D

$$R(t_R) = 1 - \Phi\left(\frac{t_R - \mu}{\sigma}\right) = R^*$$

$$1 - R^* = \Phi\left(\frac{t_R - \mu}{\sigma}\right)$$

$$t_R = \sigma \Phi^{-1}(1 - R^*) + \mu$$

b. Two-parameter Weibull

$$R(t_R) = e^{-(t/\theta)^\beta} = R^*$$

$$t_R = \theta\left[\ln\left(\frac{1}{R^*}\right)\right]^{1/\beta}$$

2. Weibull

$$E[T] = f(t) = \int_0^\infty t \frac{\beta}{\theta}\left(\frac{t}{\theta}\right)^{\beta-1} e^{-(t/\theta)^\beta} dt$$

$$u = \left(\frac{t}{\theta}\right)^\beta, \quad du = \frac{\beta}{\theta}\left(\frac{t}{\theta}\right)^{\beta-1} dt$$

Since $\Gamma(n) = \int_0^\infty x^{n-1} e^{-x} dx,$

$$E[T] = \int_0^\infty \theta u^{1/\beta} e^{-u} du = \theta \Gamma\left(1 + \frac{1}{\beta}\right)$$

a. $$E[T^2] = \int_0^\infty \theta^2 u^{2/\beta} e^{-u} du = \theta^2 \Gamma\left(1 + \frac{2}{\beta}\right)$$

$$Var[T] = \theta^2 \left\{\Gamma\left(1 + \frac{2}{\beta}\right) - \left[\Gamma\left(1 + \frac{1}{\beta}\right)\right]^2\right\}$$

b. $\beta = 2$ and $\theta = 100$ hours

$$R(t) = 1 - e^{-(t/100)^2} = 0.95$$

$$t = 100\sqrt{0.05} = 22/4$$

c. $R(200) = [1 - e^{-(200/100)^2}] = 0.9461$

4. a. $R(t|t_0) = P(T > t + t_0 | t_0) = \dfrac{R(t+t_0)}{R(t_0)}$

   From (3.6)

   $R(t) = e^{-\int_0^t \lambda(u)du}$

   $\therefore R(t|t_0) = \dfrac{e^{-\int_0^{t+t_0} \lambda(u)du}}{e^{-\int_0^{t_0} \lambda(u)du}} = e^{-\int_{t_0}^{t+t_0} \lambda(u)du}$

   b. $\lambda(t) = \dfrac{\beta}{\theta}\left(\dfrac{t}{\theta}\right)^{\beta-1}$, $\theta = 100$, $\beta = 2$

   $R(2|1) = e^{-\int_1^2 \frac{\beta}{\theta}\left(\frac{t}{\theta}\right)^{\beta-1} dt} = e^{-\int_1^2 \frac{2}{100}\left(\frac{t}{100}\right)^{\beta-1} dt} = 0.9997$

5. $R(t) = 1 - F(t) = \sum_{k=0}^{r-1} \dfrac{(\lambda t)^k}{k!} e^{-\lambda t}$

   $\lambda(t) = \dfrac{\frac{(\lambda t)^k}{k!} e^{-\lambda t}}{\sum_{k=0}^{r-1} \frac{(\lambda t)^k}{k!} e^{-\lambda t}}$

6. $f_{T_1}(t) = \lambda e^{-\lambda t}$, $f_{T_2}(t) = \dfrac{\lambda^3 t^2}{2} e^{-\lambda t}$, $t \geq 0$

   a. $T = T_1 + T_2$

   $f(t) = \int_0^t \dfrac{\lambda \tau^2}{2} \lambda e^{-\lambda \tau} e^{-\lambda(t-\tau)} d\tau = \dfrac{\lambda^4 t^3}{3!} e^{-\lambda t}$

   b. MTTF = $3/\lambda$ and IFR

   $R(t) = e^{-.05t}\left(1 + .05t + \dfrac{(.05t)^t}{2}\right)$

   c. $R(135) = .0357$, $R(150) = .0203$

   $R(150|135) = \dfrac{.0203}{.0357} = 0.5666$

7. $w = 1$, $\beta = 2$, $\theta = 3$

   $R(t) = 1 - e^{-(t/3)^2}$, $R(1) = 0.8948$

Appendix D

8. $f(t) = -\dfrac{d}{dt}R(t) = \dfrac{2}{t_0}(1 - t/t_0)$

   a. $\lambda(t) = \dfrac{2}{t_0 - t}, \quad 0 \le t \le t_0$

   b. Increasing

   c. MTTF $= t_0/3$

9. $\lambda(t) = 0.003\sqrt{t/500}, \quad t \ge 0$

   a. $R(50) = 0.96887$

   b. $R(t_d) = e^{-(8.94 \times 10^{-5})t^{-3/2}} = .9, \quad t_d = 111.8$

   c. $\dfrac{R(210)}{R(200)} \simeq 1$

10. $R_L = [1 - (1-R)^2]^2 = [2R - R^2]^2$

    $R_H = 1 - (1 - R^2)^2 = 2R^2 - R^4$

    $R_L - R_H = 2R^2(R-1)^2 \ge 0$

    $\therefore \; R_L > R_H$

11. $L \sim F(x), \; C \sim G(y)$

    $p = P(L > C) = \displaystyle\int_0^\infty P(L > y | y < C < y + dy) g_C(y) dy$

    $\therefore \; p = \displaystyle\int_0^\infty R(y) g_C(y) dy$

Chapter 4

1. $c = 25, \; \lambda = 1/MTTF = 1/30$

   $$R(w) = e^{-w/30} = .95 \;\Rightarrow\; w = 30(.0513) = 1.54$$

   Using $w = 1.5$ months,

   $$E[X(1.5)] = 25F(1.5) = 25[1 - e^{-1.5/30}] = \$1.25$$

2. $E[X(t)] = c_1 F(w)$
   a. $c_1(1-e^{-\lambda w})$
   b. $c_1\left(1 - \sum_{k=0}^{r-1} \dfrac{(\lambda w)^k}{k!} e^{-\lambda w}\right)$, using the form of the cdf in (2.8)
   c. $c_1(1 - e^{-(w/\theta)^\beta})$
3. Expected unit warranty costs
   a. Uniform (a,b)

$$F(t) = \dfrac{t-a}{b-a}, \quad a < t \le b$$

$$\therefore E[X(w)] = c_1 \left(\dfrac{w-a}{b-a}\right), \quad a < w \le b$$

   b. Normal with mean μ and variance σ^2

$$E[X(w)] = c_1 \int_{-\infty}^{w} \dfrac{1}{\sigma\sqrt{2\pi}} e^{-\left(\dfrac{t-\mu}{\sigma}\right)^2} dt$$

or

$$E[X(w)] = c_1 \Phi\left(\dfrac{w-\mu}{\sigma}\right)$$

where $\Phi(.)$ is the standard normal integral.

   c. $c_1 = \$500$, $w = 3$ years, $r = 3$, and MTTF = 4.5. Since for the Erlang $E[X(t)] = r/\lambda$, $\lambda = 3/4.5 = 2/3$ and $\lambda w = 2$.

$$E[X(w)] = 500\left[1 - \sum_{k=0}^{2} \dfrac{(\lambda w)^k}{k!} e^{-\lambda w}\right]$$

$$= 500\left[1 - e^{-2}\left(1 + \dfrac{2}{1!} + \dfrac{2^2}{2!}\right)\right] = 161.67$$

Appendix D

4. Variance of X(t)

$$E[X(t)] = c_1 F(w)$$
$$E[X(t)^2] = c_1^2 F(w)$$
$$Var[X(t)] = E[X(t)^2] - E^2[X(t)] = c_1^2[F(w) - (F(w))^2]$$

For exponential failure times

$$Var[X(t)] = c_1^2 e^{-\lambda w}(1 - e^{-\lambda w})$$

5. $E[X(w)] = c_1[\gamma(1-e^{-\lambda_1 w}) + (1-\gamma)(1-e^{-\lambda_2 w})]$
   $ = c_1[(1-e^{-\lambda_2 w}) + \gamma(e^{-\lambda_2 w} - e^{-\lambda_1 w})]$

6. $\lambda = 1/3.5$, assume a two-year economic horizon for comparison purposes
   a. Alternative A: no extended warranty
      Total Cost = product cost + maintenance cost for additional year
      $TC_A(2) = 2{,}000 + E[X(1)] = 2{,}000 + 35/3.5 = \$2{,}010$
      Alternative B: extended warranty for \$50
      Total Cost = product cost + extended warranty cost

      $$TC_B(2) = 2{,}000 + 50 = \$2{,}050$$

      Since $TC_A(2) < TC_B(2)$ do not purchase the extended warranty
   b. $TC_A(2) = \$2{,}010$ versus $TC_B(2) = 2{,}000 + x$

      $$x \begin{cases} < 10 \Rightarrow \text{choose B} \\ = 10 \Rightarrow \text{indifferent} \\ > 10 \Rightarrow \text{choose A} \end{cases}$$

7. $c = 50$, $w = 1$, $\lambda = 1/2.5 = 0.4$, $C(1) = 50(.4)(1) = \$20$
   a. For a 20 percent reduction in $C(w)$,

      $$\therefore \; C^0(w) = 50(.8) = 50\lambda(1) \;\Rightarrow\; \lambda^0 \leq 0.32$$

   b. Assume that for 3-year cost effectiveness $C^0(3) \leq C(1)$

      $$50\lambda^0(3) \leq 20 \;\Rightarrow\; \lambda^0 \leq 0.13$$

8. a. Uniform

$$E[X(w)] = c_1 \int_a^w \frac{e^{-\delta t}}{b-a} dt = \frac{c_1}{\delta(b-a)} [e^{-\delta a} - e^{-\delta w}], \quad 0 < a < w \le b$$

b. Normal

$$E[X(w)] = c_1 \int_{-\infty}^w e^{-\delta t} \frac{e^{-\left(\frac{t-\mu}{\sigma}\right)^2}}{\sigma\sqrt{2\pi}} dt = c_1 \int_{-\infty}^w \frac{e^{-\left[\left(\frac{t-\mu}{\sigma}\right)^2 + \delta t\right]}}{\sigma\sqrt{2\pi}} dt$$

Completing the square in the brackets of the exponential term and simplifying,

$$E[X(w)] = c_1 e^{(\mu\delta + \sigma^2\delta^2/2)} \int_{-\infty}^w \frac{e^{-\left(\frac{t-(\mu+\sigma\delta)}{\sigma}\right)^2}}{\sigma\sqrt{2\pi}} dt$$

11. $$E[X(w)] = \frac{c}{w_2 - w_1} \int_{w_1}^{w_2} \int_0^t e^{-\delta\tau} \lambda e^{-\lambda\tau} d\tau dt$$

$$= \frac{\lambda c}{w_2 - w_1} \int_{w_1}^{w_2} \int_0^t e^{-(\lambda+\delta)\tau} d\tau dt$$

$$= \frac{\lambda c}{\lambda + \delta} - \frac{\lambda c}{(w_2 - w_1)(\lambda + \delta)^2} [e^{-(\lambda+\delta)w_1} - e^{-(\lambda+\delta)w_2}]$$

12. $$w = F^{-1}\left\{1 - \int_{w_1}^{w_2} \frac{e^{-(t/\theta)^\beta}}{w_2 - w_1}\right\}$$

13. $$\bar{M}(s) = \left(\frac{\lambda_1 \lambda_2}{(1-\gamma)\lambda_1 + \gamma\lambda_2}\right)\frac{1}{s^3} + \left(\frac{\gamma\lambda_1 + (1-\gamma)\lambda_2}{(1-\gamma)\lambda_1 + \gamma\lambda_2}\right)\frac{1}{s^2}$$

$$M_S(w) = \frac{w}{8} - \frac{7}{32}$$

14. $$\frac{d}{dw}TC(w) = \frac{c_1}{8} + 2(25 - 2w)(-2) = 0$$

$$w = \frac{100 - 8c_1}{8}$$

Appendix D

Chapter 5

1. A. performance, reliability, conformance, (perceived quality)
   B. durability, reliability, conformance, aesthetics, (perceived quality)
   C. performance, durability, reliability, conformance, (perceived quality)

2. a. $x_1 = 0.1$, $x_2 = 45$, $x_3 = 3$

   $$q_j = \frac{x_j - u_j}{v_j - u_j}; \quad q_1 = 0.357, q_2 = 0.673, q_3 = 0.5$$

   $$\alpha_1 = \alpha_2 = \alpha_3 = 1/3; \quad \pi(q) = 0.51$$

   b. $x_1 = 0.08$, $x_2 = 20$, $x_3 = 3$

   $$q_1 = 0.5, q_2 = 0.192, q_3 = 0.5; \quad \pi(q) = 0.603$$

3. $\pi(q) = e^{\sum_j \alpha_j \ln q_j}$

   where

   $$q_j = \frac{x_j - u_j}{v_j - u_j}, \quad 0 < q_j < 1, j = 1, \ldots, n.$$

4. $\Omega_C = \begin{cases} \omega_{C1}, & x_1 < 20K \\ \omega_{C2}, & 20K \leq x_1 < 30K \\ \omega_{C3}, & 30K \leq x_1 < 40K \\ \omega_{C4}, & x_1 \geq 40K \end{cases}$, $\Omega_R = \begin{cases} \omega_{R1}, & x_2 < 0.9 \\ \omega_{R2}, & .9 \leq x_2 < 0.95 \\ \omega_{R3}, & .95 \leq x_2 < 0.99 \\ \omega_{R4}, & x_2 \geq 0.99 \end{cases}$

   $$\Omega_S = \omega_{Si} = i; \quad i = 1, \ldots, 4$$

   Set,
   $U(\omega_{C1}, \omega_{R4}, \omega_{S4}) = 1$, $U(\omega_{C4}, \omega_{R1}, \omega_{S1}) = 0$, with the remaining $0 < U(\omega_{Ci}, \omega_{Rj}, \omega_{Sk}) < 1$ are assigned according to their relative preferences between these.

5. a. Major quality elements:
   conformance quality
   reliability (MTTF)
   perceived quality (survey ranking)
   durability is also a possibility but unlikely to be significant for this pen.

b. Quality index

Variable	Attribute Level	Least Preferred	Most Preferred
Fraction nonconforming	0.04	0.10	0.01
MTTF	25	18	30
Ranking	8	12	1

$$q_1 = \frac{.04 - .10}{.01 - .10} = .67, \quad q_2 = \frac{25 - 18}{30 - 18} = .58, \quad q_3 = \frac{8 - 12}{1 - 12} = .36$$

$$\pi(q) = .67\alpha_1 + .58\alpha_2 + .36\alpha_3$$

6. $E[N_{ij}] = n_{ij} p_{ij} = \lambda_{1j}$; since $p_{1,j} = \ldots = p_{12,j}$

   a.

j:	1	2	3	4	5	6	7	8	9	10	11	12
n_{ij}	10	12	12	15	15	20	15	15	12	12	10	10
p_{ij}	.01	.015	.015	.02	.025	.015	.01	.01	.01	.01	.01	.01
$\lambda_{ij}/1000$	100	180	180	300	375	300	150	150	120	120	100	100

b. $E[N_i(6)] = \sum_{j=1}^{6} \lambda_{ij} = 100 + 180 + 180 + 300 + 375 + 300 = 1,435$

c. $E[N_i(11)] = E[N_i(6)] + \sum_{j=7}^{11} \lambda_{ij} = 2,175$

7. $E[MOP - 1, MIS - 2] = E[MOP - 1, MIS - 2 | Lag\ 0] P(Lag\ 0)$

   $+ E[MOP - 1, MIS - 2 | Lag\ 1] P(Lag\ 1) + E[MOP - 1, MIS - 2 | Lag\ 2] P(Lag\ 2)$

   $+ E[MOP - 1, MIS - 2 | Lag\ 3] P(Lag\ 3)$

   a. $\lambda_{1,1} = 100, \quad E[N_{1,1}] = 100$

   $E[NR\ Feb] = 100 P(Lag\ 0) = 70$

   $E[NR\ Mar] = 100 P(Lag\ 1) = 15$

   $E[NR\ Apr] = 100 P(Lag\ 2) = 10$

Appendix D 185

b. For the month of April

E[NR(4)] = E{# in March with MIS-1 and Lag 0}
   + E{# in February with MIS-2 and Lag 0}
   + E{# in February with MIS-1 and Lag 1}
   + E{# in January with MIS-3 and Lag 0}
   + E{# in January with MIS-2 and Lag 1}
   + E{# in January with MIS-1 and Lag 2}

$$E[NR(4)] = \lambda_{3,1}\alpha_0 + \lambda_{2,2}\alpha_0 + \lambda_{2,1}\alpha_1 + \lambda_{1,3}\alpha_0 + \lambda_{1,2}\alpha_1 + \lambda_{1,1}\alpha_2$$

8.

For MIS-1:	Volume	$m_i,1$			$D_1(t)$
MOP-t	1	10	5	10	5	0.5
	2	10	4.9	20	9.9	0.495
	3	10	4.2	30	14.1	0.47
	4	10	4	40	18.1	0.4525
	5	10	3.6	50	21.7	0.434
	6	15	3.2	65	24.9	0.383077
	7	15	2.9	80	27.8	0.3475
	8	15	3.1	95	30.9	0.325263
	9	15	3.1	110	34	0.309091
	10	15	2.8	125	36.8	0.2944
	11	10	2.7	135	39.5	0.292593
	12	10	2.6	145	42.1	0.290345
	13	10	2.6	155	44.7	0.288387
	14	10	2.5	165	47.2	0.286061
	15	10	2.3	175	49.5	0.282857

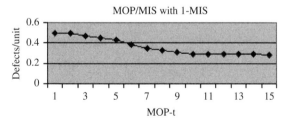

For 6-MIS:	Volume	$m_i,6$			$D_6(t)$
MOP-t.	1	10	3	10	3	0.3
	2	10	4.2	20	7.2	0.36
	3	10	3.6	30	10.8	0.36
	4	10	3.9	40	14.7	0.3675
	5	10	3.2	50	17.9	0.358

For 6-MIS:	Volume	$m_i, 6$				$D_6(t)$
MOP-t.	6	15	2.7	65	20.6	0.316923
	7	15	2.82	80	23.42	0.29275
	8	15	2.84	95	26.26	0.276421
	9	15	2.9	110	29.16	0.265091
	10	15	2.75	125	31.91	0.25528

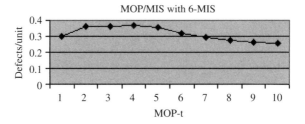

Chapter 6

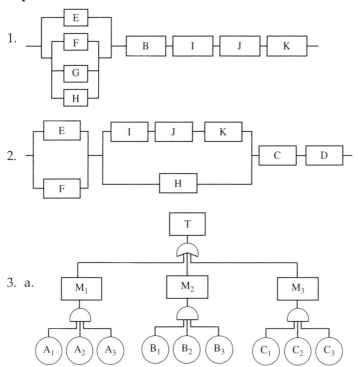

Appendix D 187

b.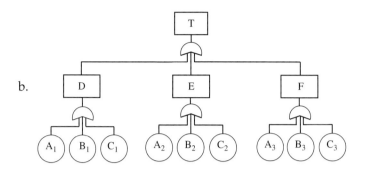

4. (a)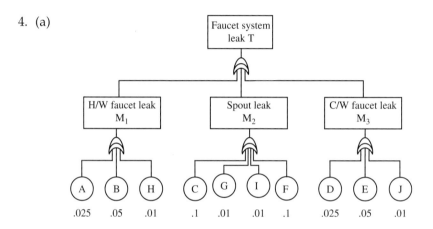

(b) $T = \{M_1 \cup M_2 \cup M_3\}$

$= \{(A \cup B \cup H) \cup (C \cup G \cup I \cup F) \cup (D \cup E \cup J)\}$

(c) $P\{T\} = (0.025 + 0.05 + 0.01) + (.1 + .01 + .01 + .1) + (0.025 + .05 + .01) = .525$

5. $T = \{B \cup C \cup D\} = \{(E \cup F) \cup (G \cap H) \cup (I \cup J)\}$

∴ $\langle E, F, (G, H), I, J \rangle$

6. Assume independence.

$$P\{T\} = P(E) + P(F) + P(G)P(H) + P(I) + P(J)$$
$$= P(E) + P(F) + P(G)P(H) + P(I) + P(J)$$
$$= .05 + .05 + (.15)(.1) + .05 + .05 = 0.215$$

7. For a 10 percent improvement, $\gamma = 0.10$.

$$R^0(1) \geq (1.1)(0.58566) = 0.64423$$

$$\lambda^0 \leq \ln\left(\frac{1}{.64423}\right) = 0.4397$$

$$\sum_1^5 c_{wi}\lambda_i T_w = 4.4647$$

$\beta_1 = .04518, \beta_2 = .18195, \beta_3 = .23803, \beta_4 = .08862, \beta_5 = .44622,$

$$\lambda_1 \leq .02737 - .04518(.53501 - .4397) = .02306$$
$$\lambda_2 \leq .07796 - .18195(.53501 - .4397) = .06062$$
$$\lambda_3 \leq .11093 - .23803(.53501 - .4397) = .08824$$
$$\lambda_4 \leq .07668 - .08862(.53501 - .4397) = .06823$$
$$\lambda_5 \leq .24207 - .44622(.53501 - .4397) = .19954$$

8. $T_w = 1, \quad \lambda = .005 + .05 + .05 + .10 = 0.205$

$$R(1) = e^{-.205(1)} = 0.81465$$

$$R^0(1) \geq (1.1)(.81465) = 0.8961$$

$$\therefore \quad \lambda^0 \leq \ln\left(\frac{1}{.8961}\right) = .10969$$

$$\sum_1^4 c_{wi}\lambda_i T_w = 15(.005) + 10(.05) + 10(.05) + 7.5(.1) = 1.825$$

$$\beta_1 = \frac{15(.005)}{1.825} = .0411, \quad \beta_2 = \beta_3 = 0.274, \quad \beta_4 = 0.411$$

Appendix D 189

$$\lambda_1^0 \le .005 - .0411(.205 - .1097) = 0.00108$$

$$\lambda_2^0 = \lambda_3^0 \le .05 - .274(.205 - .1097) = 0.02389$$

$$\lambda_4^0 \le .10 - .411(.205 - .1097) = 0.0608$$

9. $R(1) = e^{-\left[\lambda_1(1)+\lambda_2(1)+\lambda_3(1)+\frac{\lambda_4}{2}(1)^2\right]} = e^{-\left[.005+.05+.05+\frac{.1}{2}\right]} = 0.8564$

$$R^0(1) = (1.1)(.8564) = 0.9421$$

$$M_1(1) = .005, \quad M_2(1) = M_3(1) = .05, \quad M_4(1) = \frac{1}{4}e^{-2(.1)(1)} + \frac{.1}{2}(1) - \frac{1}{4} = 0.0047$$

$$\beta_1 = \frac{15(.005)}{15(.005)+10(.05)+10(.05)+7.5(.0047)} = 0.06755,$$

$$\beta_2 = \beta_3 = 0.4503, \quad \beta_4 = 0.0338$$

$$M(1) = .005 + .05 + .05 + .0047 = 0.1097$$

$$\lambda^0 = \ln\left(\frac{1}{.9421}\right) = 0.059$$

$$\lambda_1^0 = .005 - .06755(.1097 - .059) = 0.0016$$

10. (a) $EAP(a_1) = 5.6(.2) + 9(.6) + 12.5(.2) = 9.02$

$EAP(a_2) = 6.8(.2) + 7.9(.6) + 11(.2) = 8.3$

$EAP(a_3) = 5(.2) + 7.1(.6) + 14.1(.2) = 8.08$

$\max_i EAP(a_i) = 9.02 \implies a_1$

(b) a_1: $P(sales > 7.5) = P(\theta_2) + P(\theta_3) = .8$

a_2: $P(sales > 7.5) = P(\theta_2) + P(\theta_3) = .8$

a_3: $P(sales > 7.5) = P(\theta_3) = .2$

Either a_1 or a_2

(c) maximin principle

$$\min_j V(a_1, \theta_j) = V(a_1, \theta_1) = 5.6$$

$$\min_j V(a_2, \theta_j) = V(a_2, \theta_1) = 6.8$$

$$\min_j V(a_3, \theta_j) = V(a_3, \theta_1) = 5.0$$

$$\max(5.6, 6.8, 5.0) = 6.8 \quad \Rightarrow \quad \text{choose } a_2$$

11. Define the alternatives as,

a_1: copper tube system

a_2: Manguera cable

a_3: single unit design

$$\max_i EAP(a_i)$$

$EAP(a_1) = 0$

$EAP(a_2) = -25(1/4) + 50(1/2) + 85(1/4) = \$27.50$

$EAP(a_3) = -35(1/4) + 70(1/2) + 100(1/4) = \$51.25$

$$\max_i \{0, 27.50, 51.25\} = \$51.25 \quad \Rightarrow \quad \text{choose } a_3$$

Index

A

Advisory Group on Reliability of Electronics Equipment, see AGREE
Aesthetics, 36, 59
Aging process, 106–107
AGREE (Advisory Group on Reliability of Electronics Equipment), 126
Approximation to M(t), 30–32
Asymptotic M(t), 32
Average defects, 107–108
Average warranty costs, 108

B

"Bathtub curve", 44–45
Bernoulli trials, 15, 16
Binomial failure process, 27–28
Block diagram, 117
Brakes, vehicle, 123–124
Built-in redundancy, 47
"Burn-in", 45

C

Cars, 109, 110
Cartesian coordinates, 24
Cdf (cumulative distribution functions)
    Erlang distributions, 9
    failure time distributions, 8
    PRW (pro rata replacement warranty), 72
    Weibull distributions, 10
    Weibull failure times, 42
CFR (constant failure rate) allocation, 42, 45, 128–131
Chi square goodness of fit test, 25
Claims process, 95–96
Claims reporting, lags, 105
Classification, severity of failures, 118
Combined series/parallel systems, 47–50
Component reliability, 51, 127–128
Conformance, 36, 59
Conformance quality, 151
Constant failure rate, see CFR
Continuous distributions, 8–11
Continuous random variables, 14
Counting, failures, 26–32
Counting events, 11
CPUS (cost per unit sold), warranty costs, 108
Crosby philosophy, 149, 150
Cumulative distribution functions, see Cdf
Curves, product lifetime characteristic, 44
Customer behavior, 108
Cut sets, generating, 121–125

D

Decision analysis framework, 133–141
Decision consequences, 135–138
Decisions, complete uncertainty, 140–141
Decisions, risk conditions, 134–137
Decreasing failure rate, see DFR
Defects per unit sold, see DPUS
Delayed renewal processes, 25
Delays, warranty data, 109
Deming philosophy, 148–149, 150
Derivatives and integrals, 13
Design improvement, 133
DFR (decreasing failure rate), 43, 44–45
Dimensions, product quality, 36–37
Discounting and warranty costs, 62–64
Discrete distributions, 15–16
Discrete probability functions, see Dpf
Discrete random variables, 19–20
Distributed Erlangs, 68
Distribution-free methods, 22
Distributions, continuous, 8–11
Distributions, discrete, 15–16
Distributions, reliability measures, 44
Dpf (discrete probability functions)
    binomial failure processes, 28
    discrete distributions, 15
    pgf (probability generating functions), 19–20
DPUS (defects per unit sold), 107–108
Durability, 36, 59

E

Economic models for warranties, 59–80
  definitions/types, warranties, 59–60
  optimum warranty periods, determining, 74–77
  warranty cost models, 66–74
  warranty reserve, estimating, 77–80
Economy, 140
Electromechanical devices, failures, 118
Erlang distributions, 9, 16, 71, 131
Evolution, warranties, 3
Expected total warranty costs, 75
Expected warranty costs, 70–71
Exponential continuous distributions, 8
Exponential distributions, 44
Exponential failure times, 41–42, 48, 49, 78–79

F

Factorial moments, 18–19
Failure, infant mortality (examples), 45
Failure loading condition, 52
Failure mode, effect, and critical analysis, *see* FMECA
Failure probability, 51–52, 120
Failure rate and hazard function, 40–41
Failures; *see also* Product failures
Failures, electromechanical devices, 118
Failures, lifetime characteristics, 44–46
Failure time distributions, 7–21
  continuous distributions, 8–11
  discrete distributions, 15–16
  FRW (free-replacement warranty) policy, 68
  LT, 11–15
  M(t), 30
  pgf, 16–21
Failure times, 41–44, 69, 72
Fault tree analysis, *see* FTA
Fault tree structure, 121, 125
Fault tree symbols, 119
Feedback, importance of, 2
Feedback information, 150–151
FEMA (*Potential Failure Mode and Effects Analysis Manual*) block diagram, 117
Fit testing, goodness, 22–24
Fitting distributions, 21–26
FMECA (failure mode, effect, and critical analysis), 116–118
Free-replacement warranty, *see* FRW
FRW (free-replacement warranty)
  definition, 60–61
  MOP/MIS tracking, 106
  non-renewing warranties, 68–71

  renewing policy, 74
  Thomas-Richard method, 127
  unit costs, 62–63
FRW (free-replacement warranty)/PRW (pro rata replacement warranty) combined, 64–66
FTA (fault tree analysis), 116, 119–125

G

Gamma distributions, 9–10
Gamma pdf (probability functions), 15
Garvin, product quality dimensions, 150, 151
Generalized maximum entropy principle, *see* GMEP
Geometric discrete distributions, 15–16
Geometric transforms, 16–21
GMEP (generalized maximum entropy principle), 138–140
Goals, improvement, 125–133
Goodness, fit testing, 22–24

H

Hazard function, 43
Hazard rates, 131
Historical background, 2–5
History, manufacturing, 147–148
History, warranties, 59

I

Identifying problems, 116–125
IFR (increasing failure rate), 43, 45, 131–133
Improvement, reliability, 51
Improvement goals, developing, 125–133
Increasing failure rate, *see* IFR
Infant mortality failures (examples), 45
Infant mortality period, 45
Integrated product quality system, 147–152
International Standards Organization, *see* ISO
Inversion procedures, 20–21
ISO (International Standards Organization), 115, 116
ISO-9000, 116

J

Juran philosophy, 150

K

Kolmogorov-Smirnov nonparametric test, 25

L

Lag factor model, 109–111
Lags, claims reporting, 105
Laplace transform of M(t), 29
Laplace transforms, *see* LT
Lifetime characteristic curve, 44–45
Linearity, LT, 13, 19
Load, capacity models, 51–56
Load sharing, 51
Lognormal distributions, 44
LT (Laplace transforms), 11–15

M

M(t), 30–32, 67, 68
Machine failures, 23, 25–26
Manufacturing history, 147–148
Maximin principle, 140–141
Mean time to failure, *see* MTTF
MIL-HDBK 217-F, 118
Military standards, 116, 118
MIL-STD-1629A, 116
Minimax principle, 140–141
Minimum cut sets, 120, 121, 123
Minimum expected annual cost criteria, 136
Moments, Laplace transforms, 12
Month of production, *see* MOP
Month of production/months in service, *see* MOP/MIS
MOP (months of production), lags, 105
MOP/MIS (month of production/months in service) charts, 104–111, 108–109
MTTF (mean time to failure)
   exponential failure times, 48, 49
   FRW policy, 69–70
   FRW/PRW, 65
   periodic loadings, 55, 56
   reliability measures, 39–40
   renewing warranties, 74
Multiattribute quality assessment, 90–95

N

Negative binomial, discrete distributions, 16
Non-parametric methods, 22
Nonparametric test, Kolmogorov-Smirnov, 25
Nonrenewing warranty, 67–72
Normal distributions, 11, 44

O

Optimum warranty periods, determining, 74–77
Overall product quality, assessing, 91–94

P

Parallel systems, 46–47
Parametric methods, 22–26
Pascal distributions, 16
Pdf (probability density functions)
   continuous distributions, 8
   distributed Erlangs, 68
   exponential failure times, 41–42
   failure time distributions, 7
   gamma distributions, 9, 15
   IFR allocation, 131
   LT, 12, 13, 14
   normal distributions, 11
   Poisson failure process, 29
   Weibull distributions, 10
Perceived quality, 36, 59
Performance, 36, 59
Periodic loadings, 54–55
Pgf (probability generating functions), 16–21, 28
Philosophies, quality control, 148–149
Philosophy, QRW (quality, reliability and warranty), 2
Poisson events, 109
Poisson failure process, 29–30
Poisson warranty claims, 96–101
Poisson warranty claims with lags, 101–104
Potential Failure Mode and Effects Analysis Manual, *see* FMEA
Potential sources of failure, 117
Probability density, 72
Probability density functions, *see* Pdf
Probability generating functions, *see* Pgf
Probability mass, 15, 28
Probability plotting, 24–26
Product development, 37
Product failures, analyzing, 7–32
   counting, failures, 26–32
   failure time distributions (*see* Failure time distributions)
   fitting distributions, 21–26
Production and manufacturing, 37
Product life and quality, 37
Product lifetime characteristic curves, 44
Product quality, definition, 35
Product quality, warranty, 1–5, 56
Product quality dimensions, 36–37, 150
Product quality monitoring & feedback, 85–111
   MOP/MIS charts, 104–111
   multiattribute quality assessment, 90–95
   system monitoring & control, 85–90
   warranty information feedback models, 95–104

Product reliability, 38–51
  failures & life-time characteristics, 44–46
  reliability measures, 38–44
  system configurations, 46–51
Product system, 37–38
Product usage, 37
Properties, Laplace transforms, 12
Pro rata replacement warranty, see PRW
PRW (pro rata replacement warranty), 61–62, 63–64
PRW (pro rata replacement warranty) and combined policies, 72

Q

QRW (quality, reliability, and warranty), 2, 4, 150–151
QS-9000, 116
Quadratic benefit functions, 75
Quality, product, definition, 35
Quality & reliability, 35–56
  load, capacity models, 51–56
  product reliability (see Product reliability)
  quality concepts, 35–38
Quality control philosophies, 148–149
Quality improvement process, 115–141
  decision analysis framework, 133–141
  identifying problems, 116–125
  improvement goals, developing, 125–133
Quality movement, 147–150
Quality, reliability, and warranty, see QRW

R

Radar system, 49
Random failure times, 54
Random loadings, 55–56
Rebates, 61–62
Redundancy, 51
Redundancy, built-in, 47
Relative quality indicators, 94–95
Reliability, 38, 115, 127
Reliability allocation, 126
Reliability design, 115, 134
Reliability functions, 39, 54–55
Reliability improvement & redundancy, 50–51
Reliability interference, 56
Reliability measures, 38–44
Renewal counting processes, 25
Renewal density functions, 75
Renewal functions, 28, 67, 69, 131–132
Renewal processes, 25–32
Renewing warranties, 72–74
Repair/replacements, 60–61
Risk analysis, 118
Risk conditions, decisions under, 134–137

S

Safety, public, 3
Seasonal effects, 110
Series/parallel combined systems, 47–50
Series systems, 46
Simple redundancy, 47
Single-failure assumptions, 68, 69–70, 71
Standards, 116, 118
Standby redundancy, 51
States of nature, 140
Subsystems failure rates, 130
Sums of continuous random variables, 14
Sums of discrete random variables, 19–20
Symbols, fault tree, 119
System configurations, 46–51
System monitoring & control, 85–90

T

Technology, implementing new, 51
Thomas-Richard method, 127–133
Top event, 119
Total costs, 66, 67, 69–70, 75
Total product life, see TPL
TPL (total product life), 37
TPL (total product life), stages, 115
TQC systems, 88–90
Tracking, MOP/MIS, 105–106
Tracking, MOP/MIS (month of production/months in service), 105–106
Trend effects, 110

U

Uncertainty, decisions under, 140–141
Unit costs, 62–65
Usage patterns, 108
Usage stage, 37
Useful life period, 30, 45

W

Wald's equation, 67
Warranties, definitions, 59
Warranties, history, 59
Warranty burden rates, 132

Index

Warranty claims reporting, lags, 105
Warranty cost models, 66–74
Warranty costs, 70, 127, 130
Warranty costs & discounting, 62–64
Warranty data, delays, 109
Warranty information feedback models, 95–104

Warranty reserve, estimating, 77–80
Wear-out period, 45–46
Weibull distributed failure times, 30–32
Weibull distributions, 10–11, 25, 44, 69
Weibull failure times, 42–44, 79–80
Weibull slope, 42